Status of the Geoscience Workforce 2016

Carolyn Wilson

ISBN-13: 978-0-913312-54-4

Typeset in Minion Pro and Myriad Pro using Adobe InDesign CC
Graphs by Carolyn Wilson, AGI
Design and Layout by Brenna Tobler, AGI

For more information on the American Geosciences Institute and its publications, check us out at http://www.americangeosciences.org/pubs.

Front cover: University of Alaska, Anchorage Master's student surveying the tidal flats in Turnagain Arm, Alaska during his USGS internship. Photo submitted by Rob Witter, U.S. Geological Survey to AGI's 2015 Life in the Field photo contest.

Back cover photo submitted by Dan Scott to AGI's 2015 Life in the Field photo contest.

If you have any questions or comments related to this report, please contact:

Carolyn Wilson
Geoscience Workforce Data Analyst
American Geosciences Institute
4220 King Street
Alexandria, VA 22302
http://www.americangeosciences.org
Email: cwilson@americangeosciences.org
Phone: (703) 379-2480 ext. 632

american
geosciences
institute
connecting earth, science, and people

Introduction

The "Status of the Geoscience Workforce" report provides a comprehensive benchmark of the geoscience profession. This report is based on original data collected by the American Geosciences Institute and existing data from federal sources, industry sources, and professional membership organizations. This report synthesizes all the recent, readily available and reliable data related to the geosciences from the education and training of new geoscientists, to the employment trends in the geoscience workforce, to the federal funding trends for the geosciences and economic trends of the major geoscience industries. The report is broken into 5 chapters:

Chapter 1: Trends in K-12 Geoscience Education — Preparing Students for College Geoscience Programs and Society

Chapter 2: Trends in Two-Year College Geoscience Programs

Chapter 3: Trends in Geoscience Education at Four-Year Institutions

Chapter 4: Geoscience Employment Sectors — Trends in Student Transitions and Workforce Dynamics

Chapter 5: Economic Metrics and Drivers of the Geoscience Workforce

Some highlights of this report include:

- 16% of doctoral graduates, 23% of master's graduates, and 33% of bachelor's graduates had transferred from a two-year college during their education, which is an increase across all degree levels since 2013. The growth in participation at two-year colleges continues to highlight the importance of collaborations between two-year colleges and four-year institutions to help with the transition of these students into bachelor's programs.

- Field camps may have reached capacity. In 2014, attendance reached its peak at 3,237 students and decreased in 2015 to 2,867. The availability of field camps must be taken into consideration if employers are expecting students to have this skills development experience.

- 64% of bachelor's graduates, 41% of master's graduates, and 57% of doctoral graduates did not participate in an internship. However, these experiences are seen by employers as essential for career preparation and professional skills development. Those graduates that did participate in internship experiences recognized the importance of these experiences for their professional and academic development.

- With very little change in the enrollments in master's and doctoral geoscience programs, it appears these programs have reached capacity. Students seem to have recognized the competitiveness for graduate slots in the geosciences as fewer bachelor's and master's graduates indicated plans to immediately enroll in a graduate program.

- While there has been a downturn in the oil and gas industry starting at the end of 2014, the oil and gas industry was still hiring geoscience graduates, particularly with master's degrees. However, for bachelor's graduates, the environmental industry appears to be a viable job option at graduation with 40% of bachelor's graduates finding a job within this industry in 2015. Other industries, such as the nonprofit sector, have been hiring recent graduates at a higher rate than seen previously.

- While the full salary ranges for each degree level can be wide depending on the job, typically bachelor's graduates have starting salaries between $30,000-$50,000; master's graduates have starting salaries either between $40,000-$60,000 or between $100,000-$120,000; and doctoral graduates typically have starting salaries between $50,000-$70,000. Every graduate making more than $90,000 was working in the oil and gas industry.

- Median salaries for geoscientists continue to increase in all major industries. In a shift from the 2014 report, engineering managers, instead of petroleum engineers, have the highest median salary of $132,800; petroleum engineers have a median annual salary of $129,990; geoscientists have a median annual salary of $89,700; mining and geological engineers have a median annual salary of over $94,040; and hydrologists have a median annual salary of nearly $79,550.

- There were approximately 324,000 geoscientists employed in the United States in 2014. Over the next decade, 48% of the workforce will be at or near retirement. However, while there will still be a predicted shortage of around 90,000 geoscientists, this shortage continues to decrease from prior estimates because of an increase in worker productivity and the surge of new students entering the job market over the last decade.

Acknowledgements

I would like to thank the AGI Foundation for funding the AGI Workforce Program and for their support for this report.

I also want to thank the people and organizations that allowed my use of their data for this report, including: Ann Benbow and Ed Robeck for the K-12 earth science education data; Dr. Penelope Morton for the field camp attendance data; Dr. Heather Macdonald for her list of two-year colleges with geoscience programs; the National Mining Association for the data on mining employment demographics; the American Association of Petroleum Geologists, the National Ground Water Association, the Society of Economic Geologists, and the Society of Exploration Geophysicists for their membership data; The National Science Foundation's center for Science and Engineering Statistics for granting access to some of their restricted-use data files; and GeoRef Information Services for data on publications, theses, and dissertations.

I want to acknowledge the multiple organizations and agencies that freely provide their data and information online, including: the American Geophysical Union, the American Meteorological Society, the Association for the Sciences of Limnology and Oceanography, ACT, Baker Hughes, the College Board, the Energy Information Administration, the National Association of Geoscience Teachers, the National Science Foundation's National Center for Science and Engineering Statistics, the National Science Foundation award database, the U.S. Bureau of Economic Analysis, the U.S. Census Bureau, the U.S. Bureau of labor Statistics, the U.S. Department of Education's Integrated Postsecondary Education Database, the U.S. Department of Education's National Center of Education Statistics, the U.S. Geological Survey, the U.S. Office of Personnel Management, and the World Gold Council.

I want to give recognition to the AGI Workforce Program data interns over the past few years, Sebastian Corrochano and Jordan Ellington, for their hard work quality controlling the Directory of Geoscience Departments database and the raw data from AGI's Geoscience Student Exit Survey.

Contents

Tables

Figures

Chapter 1: Trends in K–12 Geoscience Education — Preparing Students for College Geoscience Programs and Society

Earth Science education is becoming more pervasive in middle school and high school curriculums across the U.S. because of an increased awareness of the importance of earth science and environmental science to a functioning society.[1] Nearly every state in the United States allows for earth science courses to count as a science credit for graduation and assesses students' knowledge of earth science concepts during middle school and high school. The College Board recognizes Earth Science as an essential domain for college readiness in science.[2] A majority of colleges also accept Earth Science courses as a creditable science course with over 77% accepting these courses for student admission into their programs.[3]

However, the earth sciences still have the fewest trained teachers at the elementary and secondary levels than other sciences. In fact, in elementary schools the majority of teachers providing computer, math and science education have degrees in the social or related sciences and non-science and engineering related fields, instead of a science or engineering field. In secondary schools, approximately half of computer, math, and science teachers have degrees in science and engineering related fields, such as health fields, and non-science and engineering related fields, such as humanities fields, instead of holding a degree in a science or engineering field.

College readiness among students is often indicated through course selection of higher level math and science courses, such as Advanced Placement (AP) course selection, and SAT test scores. There continues to be growth in the percentages of high school graduates taking higher level math courses with 75% completing Algebra II, 35% completing Pre-Calculus, and 16% completing Calculus in 2009. There is similar growth among high school graduates taking science courses, with 96% completing Biology, 70% completing Chemistry, 36% completing Physics, and 28% completing Geology/Earth Science in 2009. Science and Calculus Advanced Placement courses show similar growth over time.

Generally SAT test-takers with coursework in the sciences tend to have higher average scores than the average of all test-takers, however, this is not the case for students with coursework in the earth and space sciences. This tends to be the case because many students that take Geology or Earth Science in high school are generally not as interested in science and count it as a science elective required for graduation. Over the past few years, it appeared that SAT test-takers interested in degrees in the physical sciences has hovered just above 20,000 students, but there has been an increase in the number of students interested in interdisciplinary studies. As Earth Science tends to be a highly interdisciplinary degree area, this increase can be encouraging for the future recruitment of geoscience majors from the introductory geoscience courses offered at postsecondary degree institutions.

Since 2013, AGI surveyed students graduating with a degree in the geosciences using AGI's Geoscience Student Exit Survey about their educational background in the geosciences. Consistently over the past three years, approximately half of the graduates had taken a formal earth science, environmental science, or geography course while in high school. This indicates the usefulness of these courses to start recruiting students into geoscience majors during their postsecondary education, as well as reiterates the increasing inclusion of earth science courses in the high school curriculum.

While Earth Science education is typically presented as a single course in middle school or high school for most students, the National Research Council, the National Science Teachers Association, the American Association for the Advancement of Science, and Achieve worked together to create more integrated K-12 science curriculum with the Next Generation Science Standards in 2013. One of the recognized Disciplinary Core Ideas is Earth and Space Science, which makes earth science concepts equally represented with the other science disciplines within the state standards. Curriculum continues to be developed to integrate all areas of science throughout the entire K-12 science curriculum as set standards for what students should know at each grade level. Currently, eleven states, plus the District of Columbia, have adopted the Next Generation Science Standards, with more states looking to adopt in the near future. For more information about the Next Generation Science Standards, please visit http://www.nextgenscience.org.

[1] National Research Council, Board on Science Education. (2012) A Framework for K-12 Science Education: Practices, Crosscutting Concepts, and Core Ideas. Retrieved from http://www.nap.edu/catalog.php?record_id=13165#

[2] College Board. (2010) Science College Board Standards for College Success. Retrieved from http://media.collegeboard.com/digitalServices/pdf/research/Science_College_Board_Standards_for_College_Success_SCAS.pdf

[3] Center for Geoscience Education and Public Understanding. (2013) Earth and Space Science Education In U.S. Secondary Schools: Key Indicators and Trends. Retrieved from http://www.americangeosciences.org/sites/default/files/education-ESS-sec-status-report-2013-09-01-13.pdf

Earth Science Education

Figures 1.1 and 1.2 and Tables 1.1 and 1.2 show the increased representation of earth science education in middle and high school curriculum. One more state since 2010, Nebraska, has made earth science a required course for graduation, and 46 states in the United States will accept an earth science course as a science credit for graduation, compared to 12 states in 2010. However, it is important to note that, in 2010, all states that didn't include earth science courses toward graduation did list earth science in the state high school standards. The framework was in place then to move towards this increase in the inclusion of earth science courses as a graduation credit. Forty-eight states also include the assessment in either middle or high school.

Figure 1.1: Earth Science Education Graduation Requirements in High School

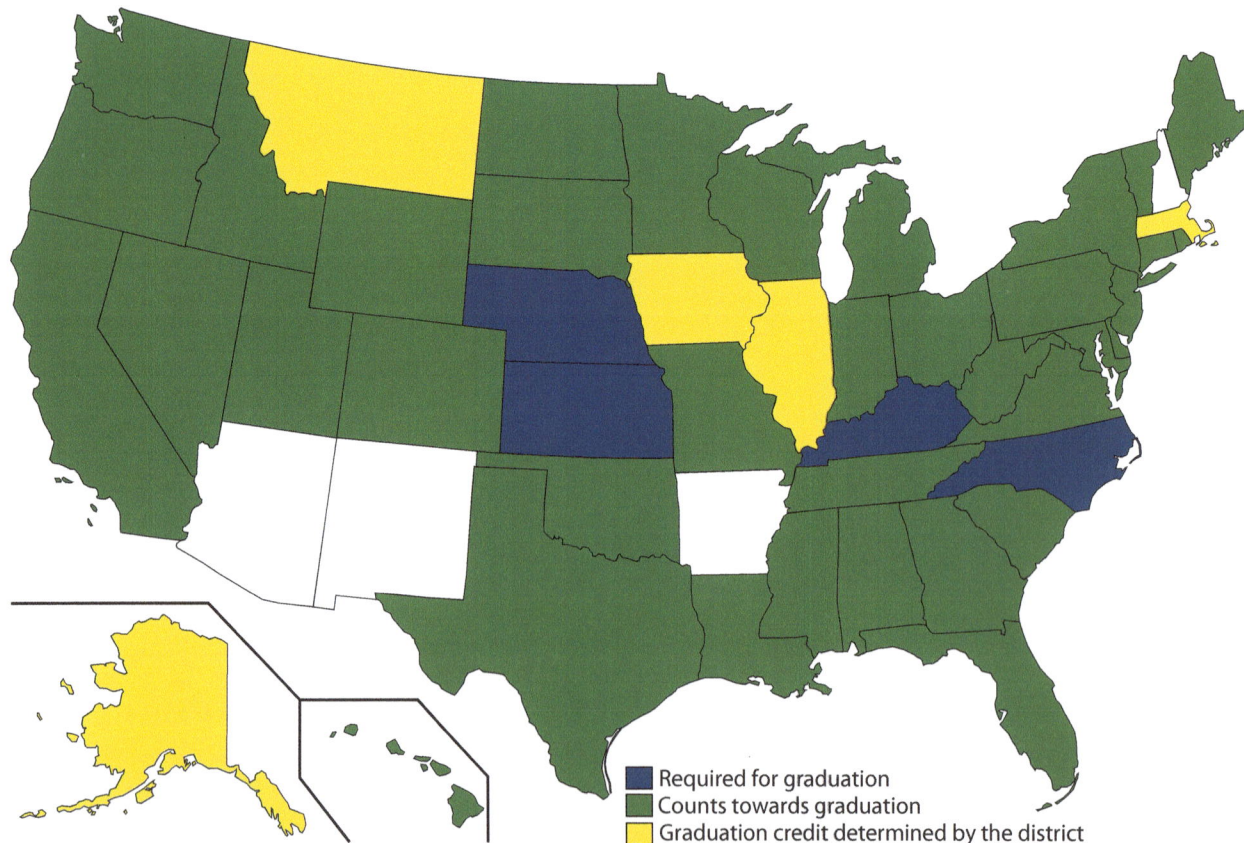

Legend:
- Required for graduation
- Counts towards graduation
- Graduation credit determined by the district

AGI Geoscience Workforce Program; Data derived from Ann Benbow and AGI's Education Program data

Table 1.1: Changes in State-Level Earth Science Requirements for Graduation

Is Earth Science a Required Course for Graduation?				
State	2002	2007	2010	2013
Alaska	No	Determined by District	No	No
Arizona	No	Determined by District	No	No
Colorado	No	Determined by District	No	No
Connecticut	No	Determined by District	No	No
Idaho	No	Yes	No	No
Illinois	No	Determined by District	No	No
Indiana	No	Yes	No	No
Kansas	No	Yes	Yes	Yes
Kentucky	Yes	Yes	Yes	Yes
Louisiana	No	Yes	No	No
Massachusetts	No	Determined by District	No	No
Michigan	No	Yes	No	No
Nebraska	No	No	No	Yes
Nevada	No	Determined by District	No	No
New Hampshire	No	Determined by District	No	No
New Jersey	No	Determined by District	No	Environmental Lab-Based Course Required
New York	Yes	No	No	No
North Carolina	Yes	Yes	Yes	Yes
North Dakota	No	Determined by District	No	No
Oregon	No	Determined by District	No	No
Pennsylvania	Yes	Determined by District	No	No
Rhode Island	No	Determined by District	No	No
Wyoming	Yes	Determined by District	No	No

AGI Geoscience Workforce Program; Data derived from Ann Benbow and AGI's Education Program data

Table 1.2: States Counting Earth Science Courses Towards Graduation Requirements

Does an Earth Science Course Count Towards Graduation Requirements?				
State	2002	2007	2010	2013
Alabama	Yes	Yes	Yes	Yes
Alaska	Yes	Determined by District	No*	Determined by District
Arizona	Yes	Determined by District	No*	Integrated Concepts into other Courses
Arkansas	Yes	No	Yes	No
California	Yes	Yes	No*	Yes
Colorado	Yes	Determined by District	No*	Yes
Connecticut	-	Yes	No*	Yes
Delaware	-	Yes	No*	Yes
District of Columbia	-	Yes	No*	Yes
Florida	Yes	Yes	Yes	Yes
Georgia	Yes	Yes	No*	Yes
Hawaii	Yes	-	No*	Yes
Idaho	Yes	Yes	No*	Yes
Illinois	Yes	Determined by District	No*	Determined by District
Indiana	Yes	Yes	Yes	Yes
Iowa	-	Determined by District	No*	Determined by District
Kansas	Yes	Yes	No*	Yes
Kentucky	Yes	Yes	No*	Yes
Louisiana	Yes	Yes	Yes	Yes
Maine	Yes	Yes	No*	Yes
Maryland	Yes	Yes	Yes	Yes
Massachusetts	Yes	Determined by District	No*	Determined by District
Michigan	No	Determined by District	No*	Yes
Minnesota	Yes	Yes	No*	Yes
Mississippi	Yes	Yes	Yes	Yes
Missouri	Yes	Yes	No*	Yes

Does an Earth Science Course Count Towards Graduation Requirements?				
State	2002	2007	2010	2013
Montana	Yes	Yes	No*	Determined by District
Nebraska	-	Determined by District	Yes	Yes
Nevada	Yes	Yes	No*	Yes
New Hampshire	No	Determined by District	No*	No
New Jersey	Yes	Yes	No*	Yes
New Mexico	Yes	Yes	No*	No
New York	Yes	Yes	Yes	Yes
North Carolina	Yes	Yes	No*	Yes
North Dakota	Yes	Determined by District	No*	Yes
Ohio	Yes	Determined by District	No*	Yes
Oklahoma	Yes	Yes	Yes	Yes
Oregon	Yes	Determined by District	No*	Yes
Pennsylvania	Yes	Yes	No*	Yes
Rhode Island	-	Determined by District	No*	Yes
South Carolina	No	Yes	No*	Yes
South Dakota	Yes	Yes, with Lab only	No*	Yes
Tennessee	-	Yes	No*	Yes
Texas	No	No	No*	Yes
Utah	Yes	Yes	Yes	Yes
Vermont	Yes	Yes	No*	Yes
Virginia	Yes	Yes	Yes	Yes
Washington	Yes	Yes, with Lab only	No*	Yes
West Virginia	Yes	Yes	No*	Yes
Wisconsin	Yes	Determined by District	No*	Yes
Wyoming	Yes	Determined by District	No*	Yes

* denotes earth science is included in the state high school science standards.

AGI Geoscience Workforce Program; Data derived from Ann Benbow and AGI's Education Program data

Figure 1.2: States that Assess Earth and Space Science Concepts in Middle and High School

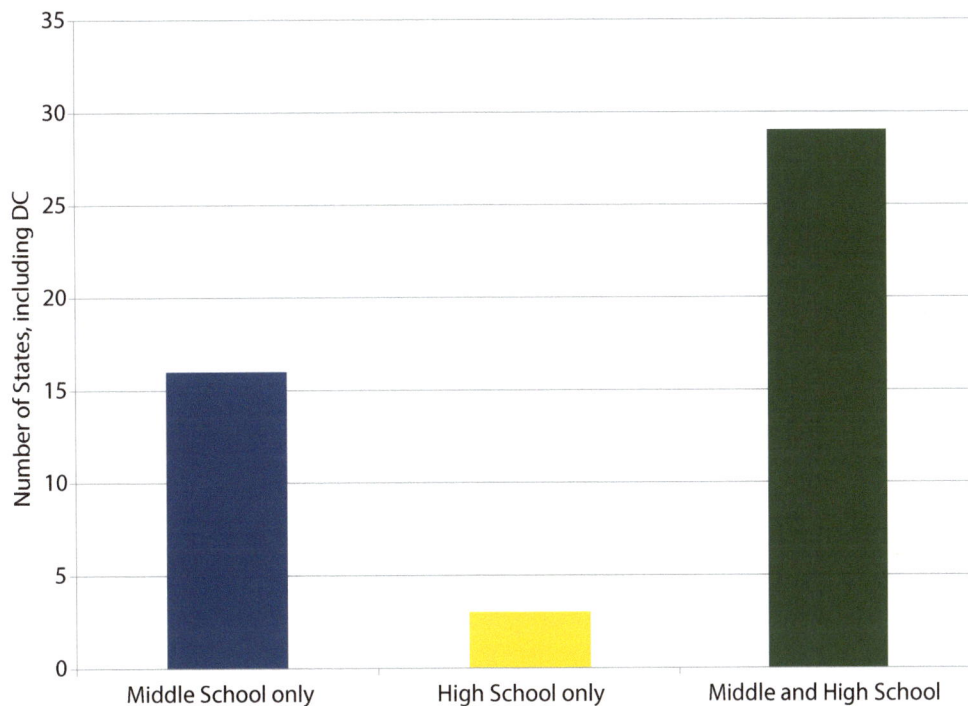

AGI Geoscience Workforce Program; Data derived from Ann Benbow and AGI's Education Program data

Teachers

The percentage of teachers presented in Figures 1.3 and 1.4 are teachers of computer science, math or science by the National Science Foundation. This year's report uses percentages instead of actual counts, as in previous editions of this report, because there was some concern over the accuracy of the NSF-applied weights to the survey data when selecting for the degree fields within the geosciences. The weights were calculated for understanding the general overall population, but as the selection of the data becomes more specific, the factors do not scale down effectively to provide confident estimates of the population—in this case the percentage of geoscience graduates in the K-12 profession.

The majority of elementary teachers hold degrees in social and related fields and non-science and engineering fields, such as the humanities. This is because most elementary teachers tend to teach multiple subjects to their students, and most have a degree in elementary education. Approximately half of the secondary teachers teaching computer science, math, or science hold degrees in non-science and engineering related fields and science and engineering related fields, such as health. Thirty-nine percent of them hold a degree in computer science, mathematics, life sciences, or physical sciences. Approximately 1% of elementary teachers and 3% of secondary teachers hold a geoscience degree.

Figure 1.3: Degree Fields of Elementary School Teachers, 2013

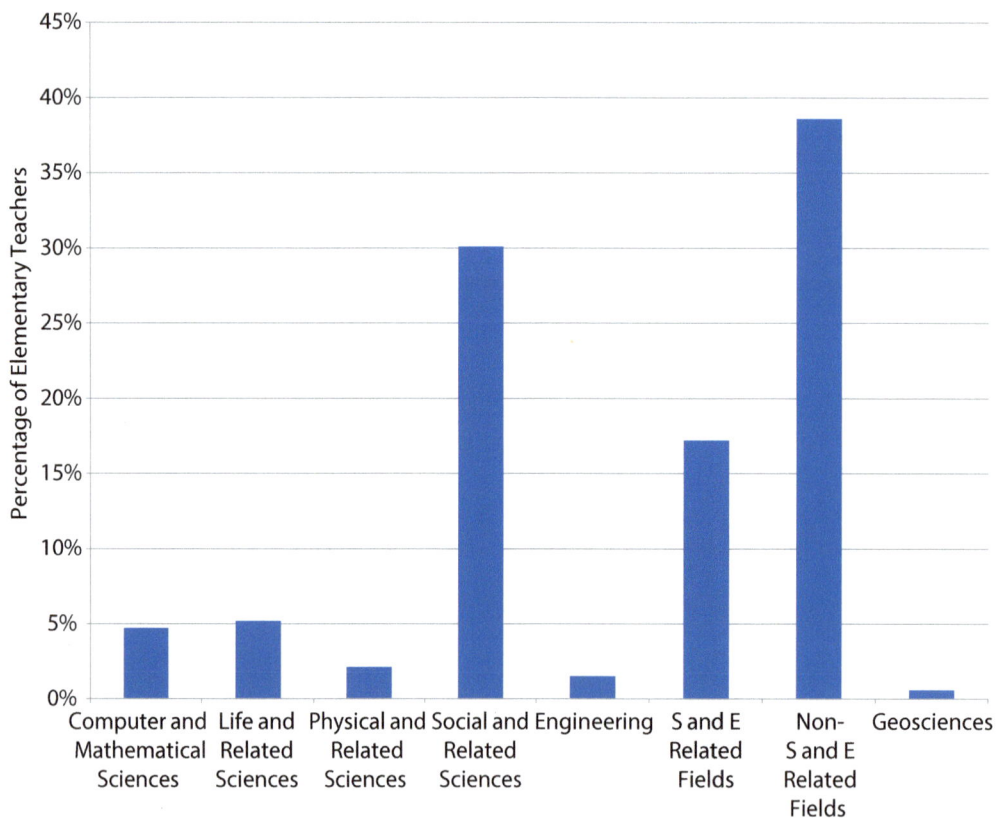

AGI Geoscience Workforce Program, Data derived from NSF's SESTAT Restricted-Use data files

Figure 1.4: Degree Fields of STEM Secondary Teachers, 2013

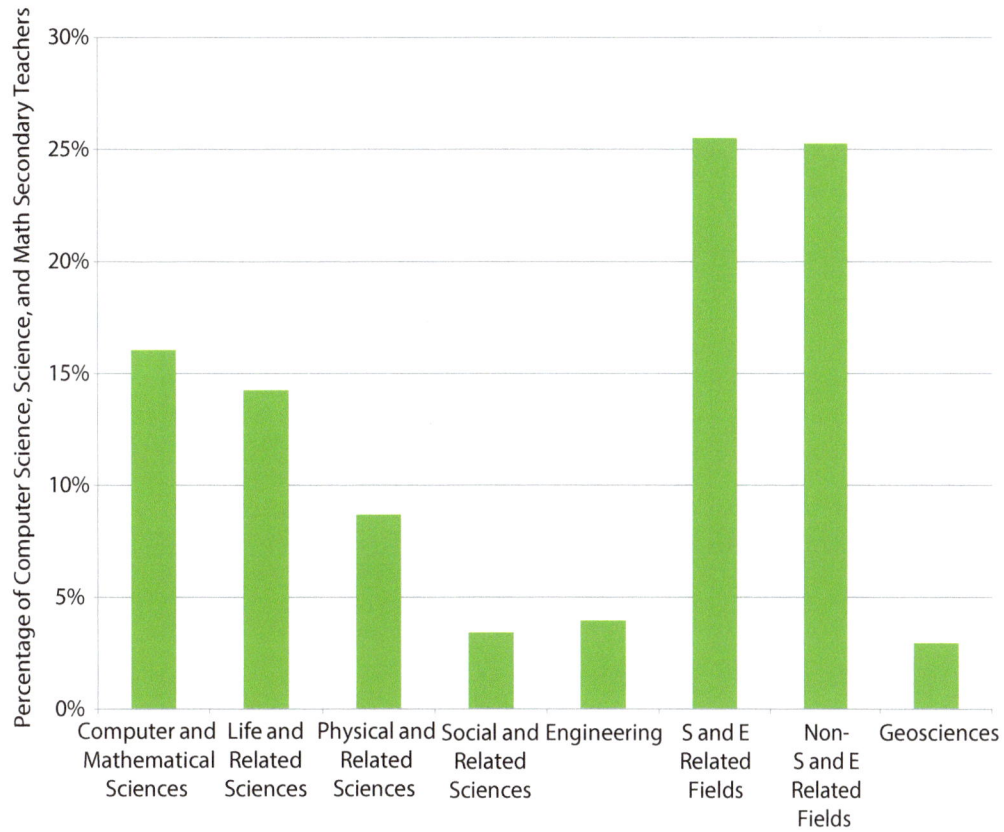

AGI Geoscience Workforce Program; Data derived from NSF's SESTAT Restricted-Use data files

Students

High school course selection in math and science and Advanced Placement (AP) course selection are both used as indicators of future college success by demonstrating students' ability to handle the rigorous college coursework. Continually increasing percentages of students are completing coursework in all math courses presented in Figure 1.7, except trigonometry. The same is generally true for the science courses in Figure 1.10. While it appears the percentages of graduates taking geology/earth science varied some for 12 years, it appears this subject is steadily increasing in the percentage of students since 2000.

Over the past decade, Advanced Placement course participation in science and Calculus has steadily increased in the United States. While Biology, Chemistry, and Physics tend to be more popular AP courses than Environmental Sciences, but the Environmental Sciences AP course is still relatively new. The growth in this course is encouraging for the recruitment of future geoscience students

Figures 1.14-1.17 present data about SAT test-takers with coursework in the sciences. Average SAT scores in critical reading, writing, and math appear low for students that took geology, earth, or space science. However, it is important to note that this course is typically considered

a science elective for graduation in the United States, so it can tend to draw lower achieving students and those students not interested in offered AP courses. In general, though, more SAT test-takers have been taking coursework in chemistry and physics in 2015 compared to test-takers in 2000, but there is a decreasing trend in average math, critical reading and writing scores on the SAT over the past ten years among students with coursework in science. This decrease in SAT scores seems to be corresponding with an increase in the number of students that want a postsecondary degree, particularly a bachelor's degree, with an increase from 264,000 students in 2006 to 441,000 students in 2015 (Figure 1.18). Along with the decreases in SAT scores, the ACT program has noted decreases in the percentage of high school students reaching the college readiness benchmarks in reading and English. However, they are seeing an increase in the college readiness in science from 27% in 2006 to 38% in 2015.

AGI's Geoscience Student Exit Survey has revealed approximately half of the graduates at all degree levels took a formal earth science course while in high school (Figure 1.21). These results have been consistent over the past three years. This indicates the importance of these high school courses for recruitment of future geoscience majors at the postsecondary level.

Figure 1.5: Selected Math Courses Taken by U.S. High School Graduates

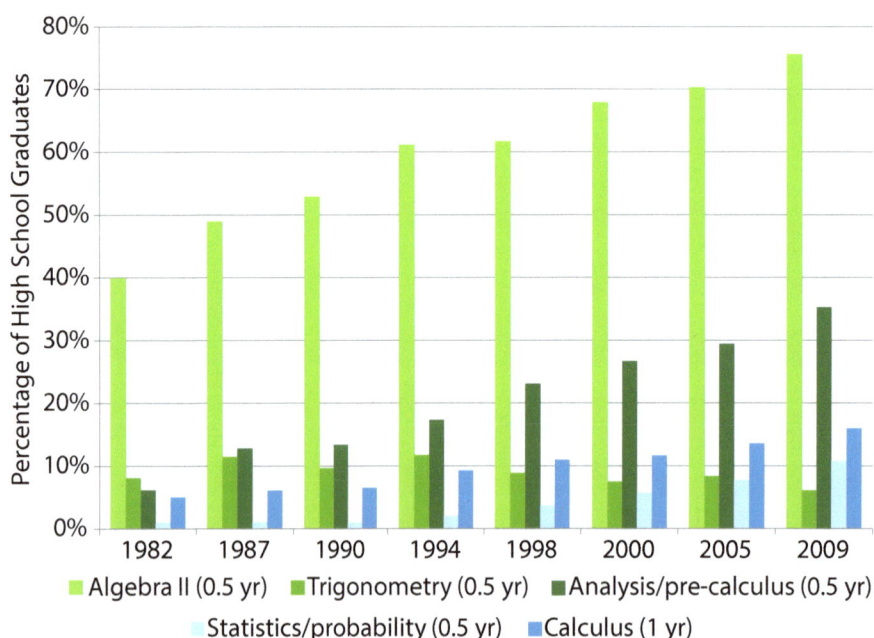

AGI Geoscience Workforce Program; Data derived from NCES Digest of Education Statistics, 2012

Figure 1.6: Selected Math Courses Taken by Gender of U.S. High School Graduates, 2009

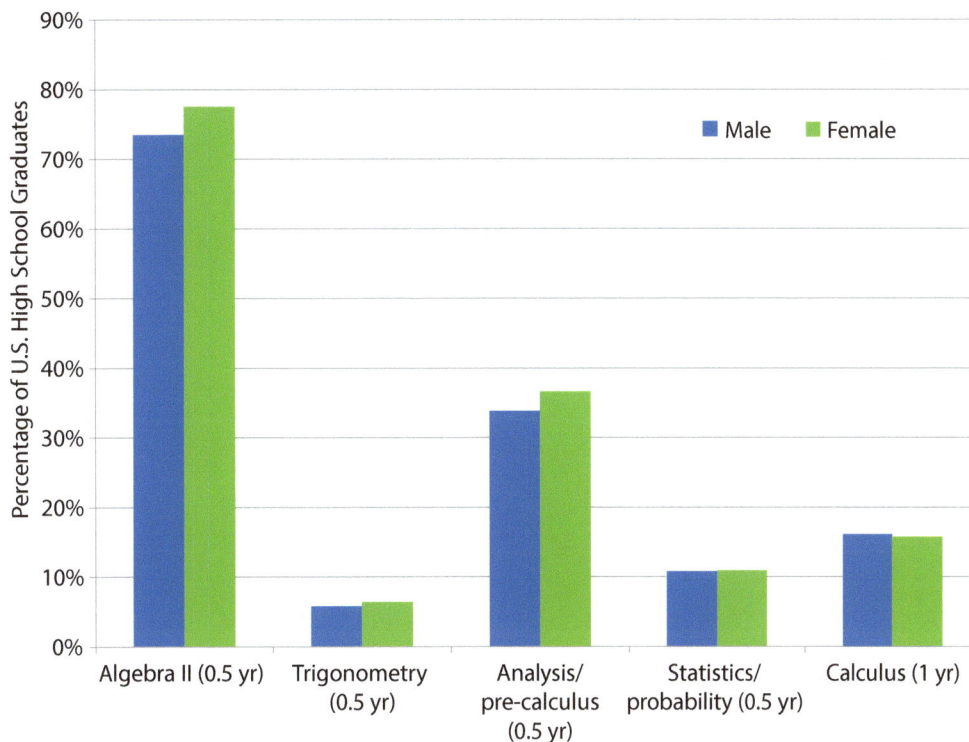

AGI Geoscience Workforce Program; Data derived from NCES Digest of Education Statistics, 2012

Figure 1.7: Selected Math Courses Taken by Race and Ethnicity of U.S. High School Graduates, 2009

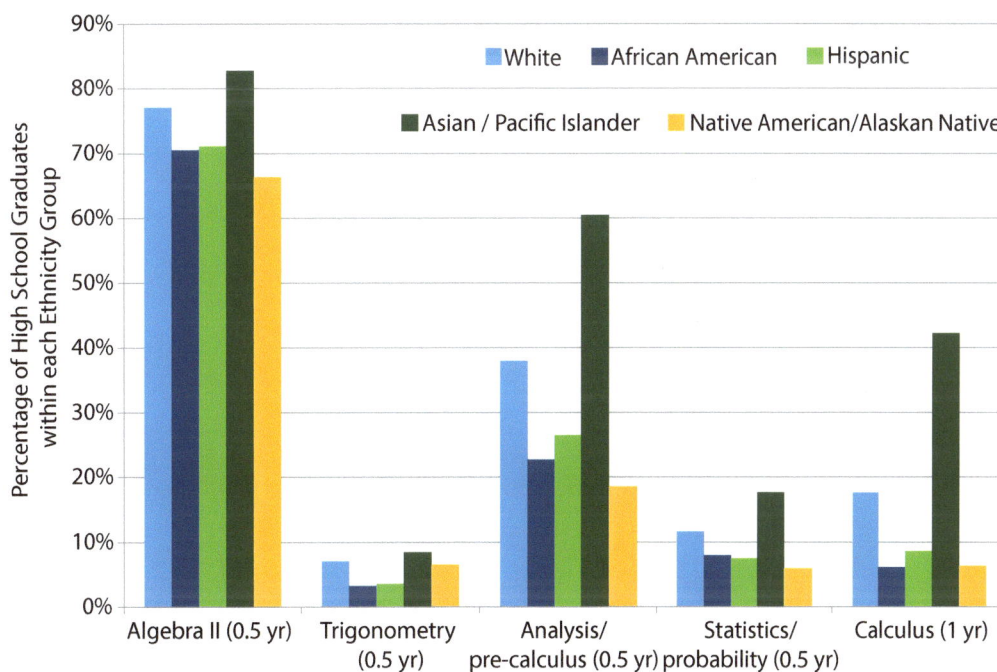

AGI Geoscience Workforce Program; Data derived from NCES Digest of Education Statistics, 2012

Figure 1.8: Science Courses Taken by U.S. High School Graduates

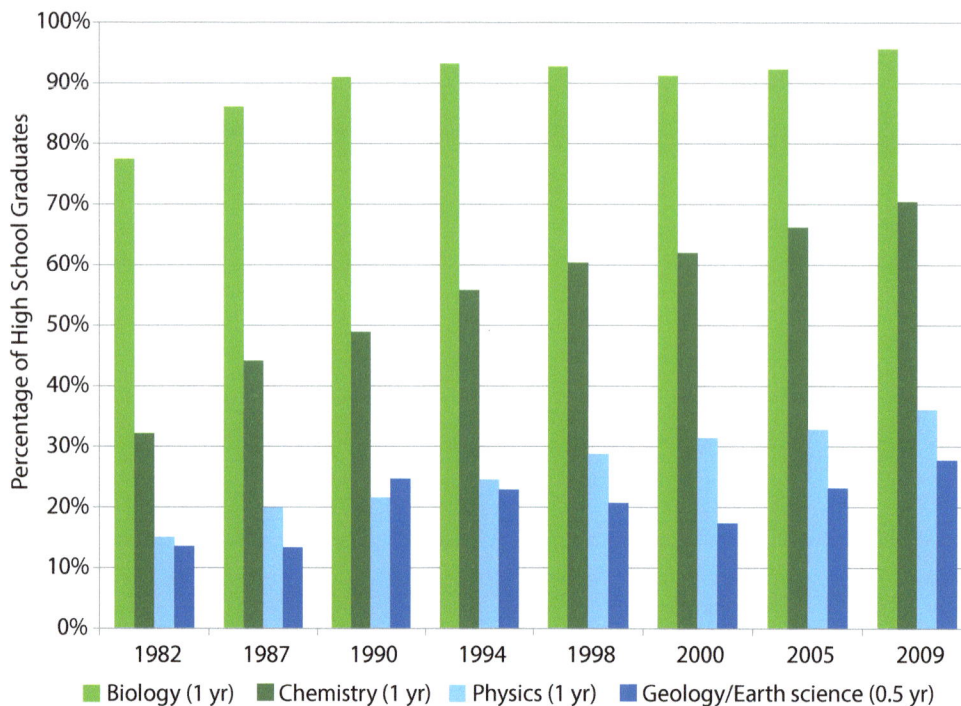

AGI Geoscience Workforce Program; Data derived from NCES Digest of Education Statistics, 2012

Figure 1.9: Science Courses Taken by Gender of U.S. High School Graduates, 2009

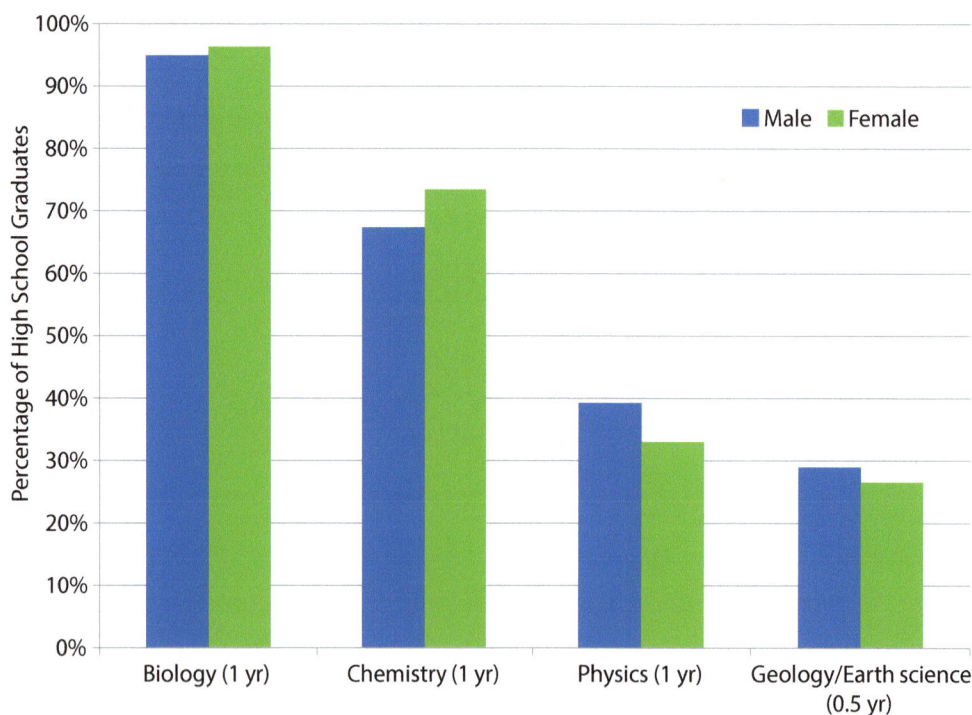

AGI Geoscience Workforce Program; Data derived from NCES Digest of Education Statistics, 2012

Figure 1.10: Science Courses Taken by Race and Ethnicity of U.S. High School Graduates, 2009

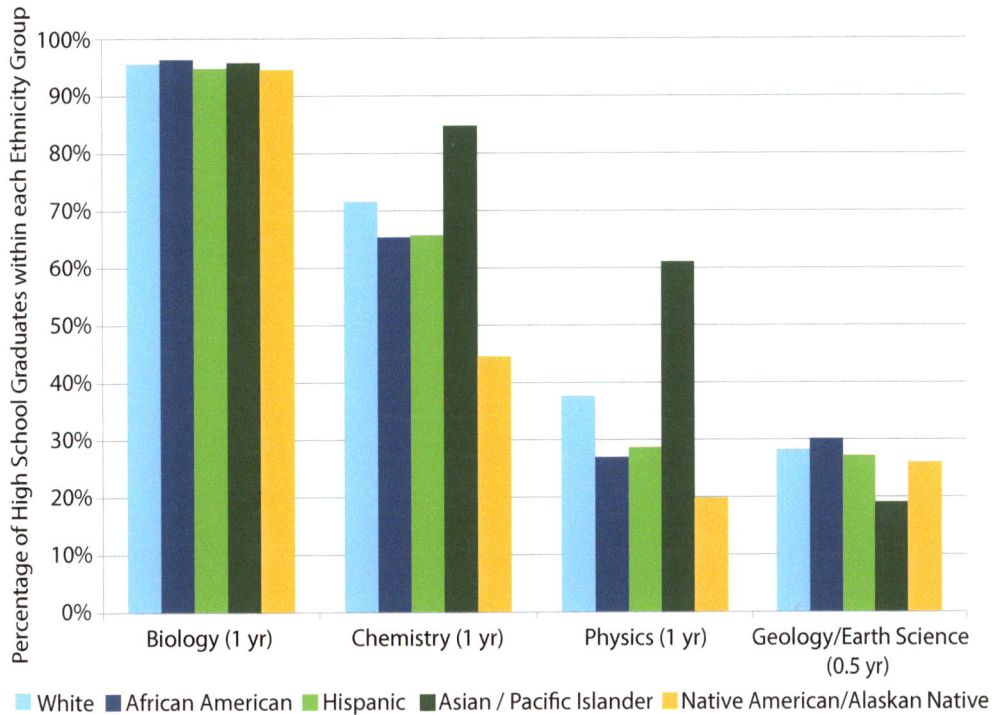

White ■ African American ■ Hispanic ■ Asian / Pacific Islander ■ Native American/Alaskan Native

AGI Geoscience Workforce Program; Data derived from NCES Digest of Education Statistics, 2012

Figure 1.11: Number of Students Taking AP Courses in Science and Math, 2006-2015

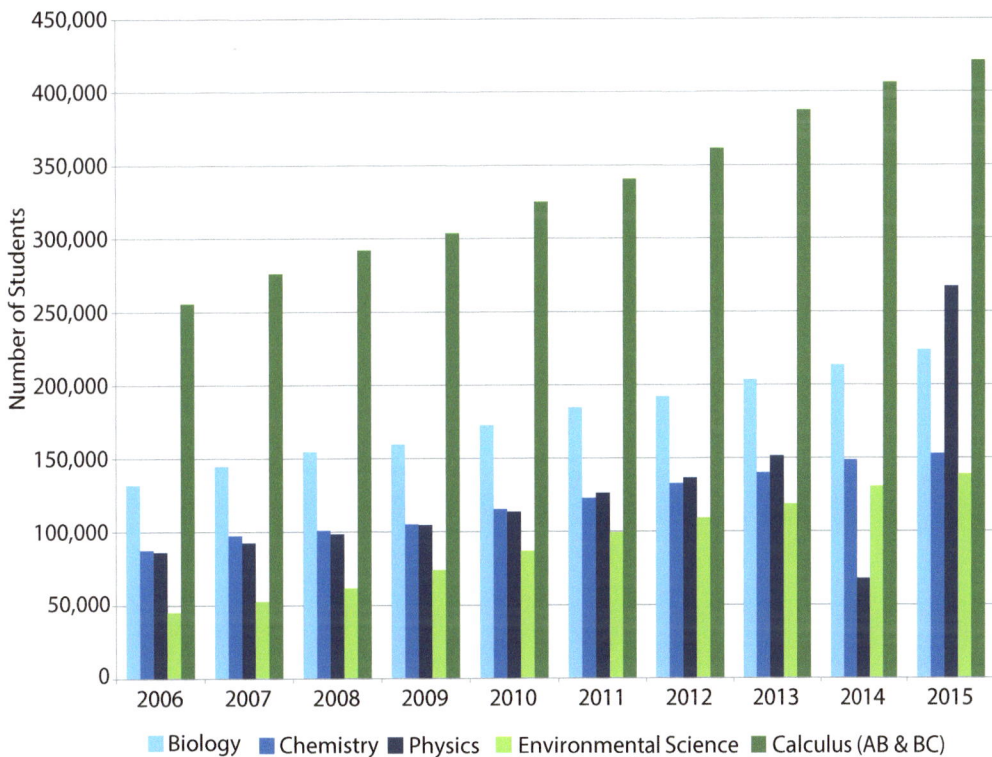

■ Biology ■ Chemistry ■ Physics ■ Environmental Science ■ Calculus (AB & BC)

AGI Geoscience Workforce Program; Data derived from the College Board AP Data

Figure 1.12: Advanced Placement Courses Taken by Gender of U.S. High School Graduates, 2009

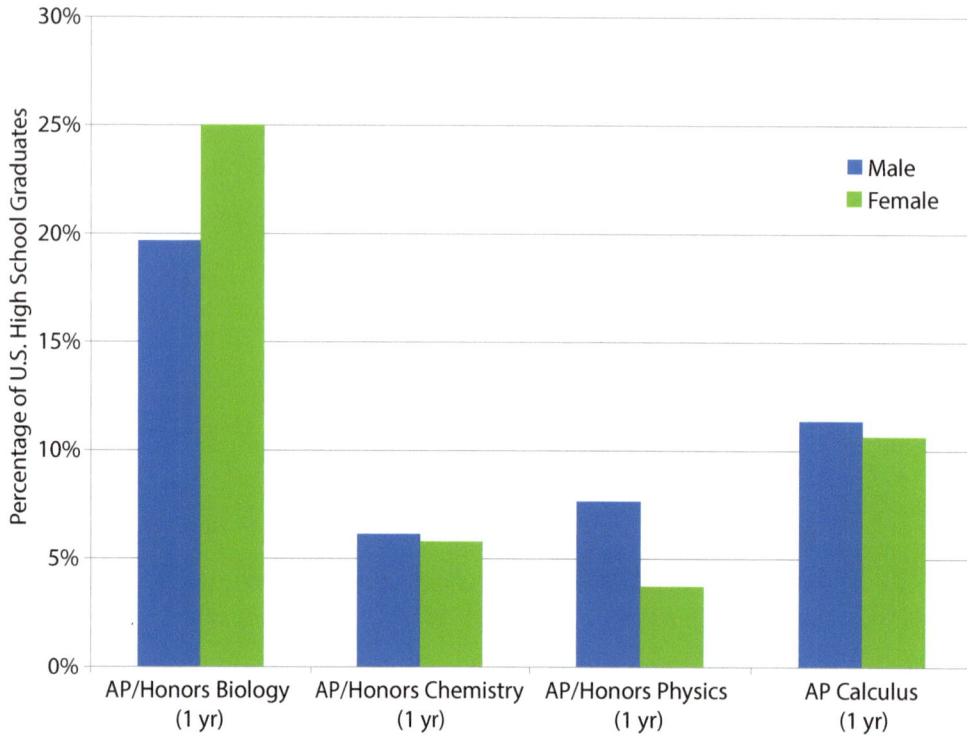

AGI Geoscience Workforce Program; Data derived from NCES Digest of Education Statistics, 2012

Figure 1.13: Advanced Placement Courses Taken by Race and Ethnicity of U.S. High School Students, 2009

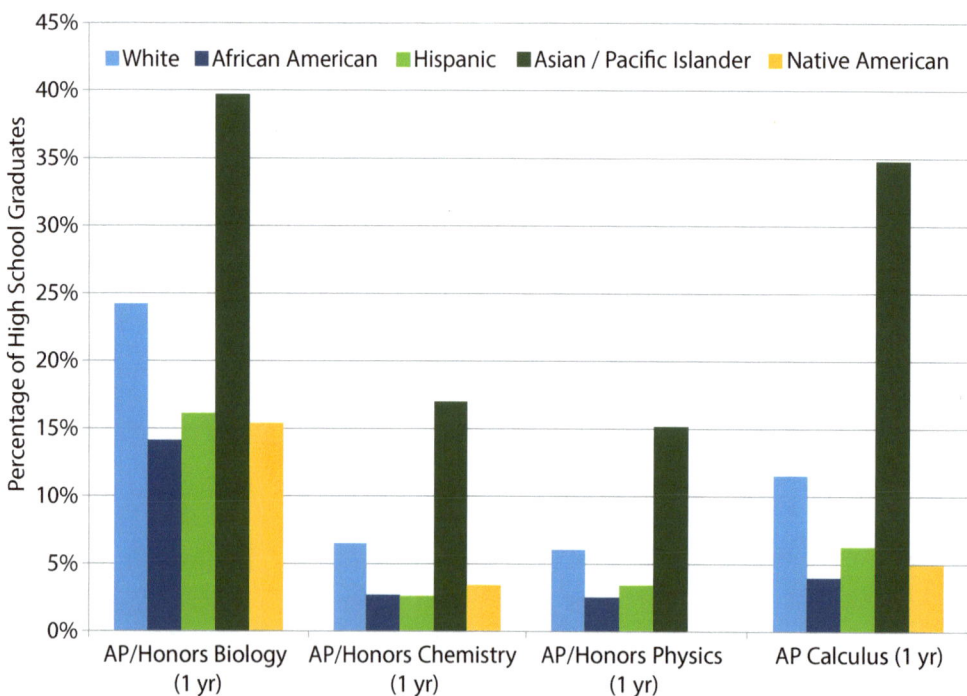

AGI Geoscience Workforce Program; Data derived from NCES Digest of Education Statistics, 2012

Figure 1.14: SAT Test-Takers with Coursework or Experience in Selected Sciences

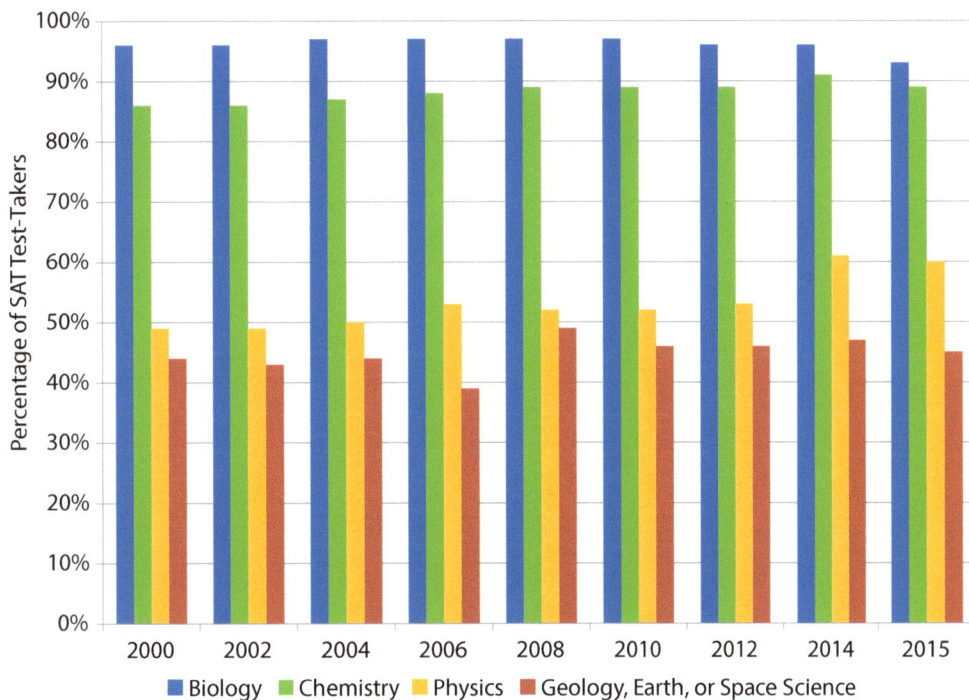

AGI Geoscience Workforce Program; Data derived from the College Board College-Bound Seniors, Total Group Report, 1996-2015

Figure 1.15: Mean Math SAT Scores for Test-Takers with Coursework in Science

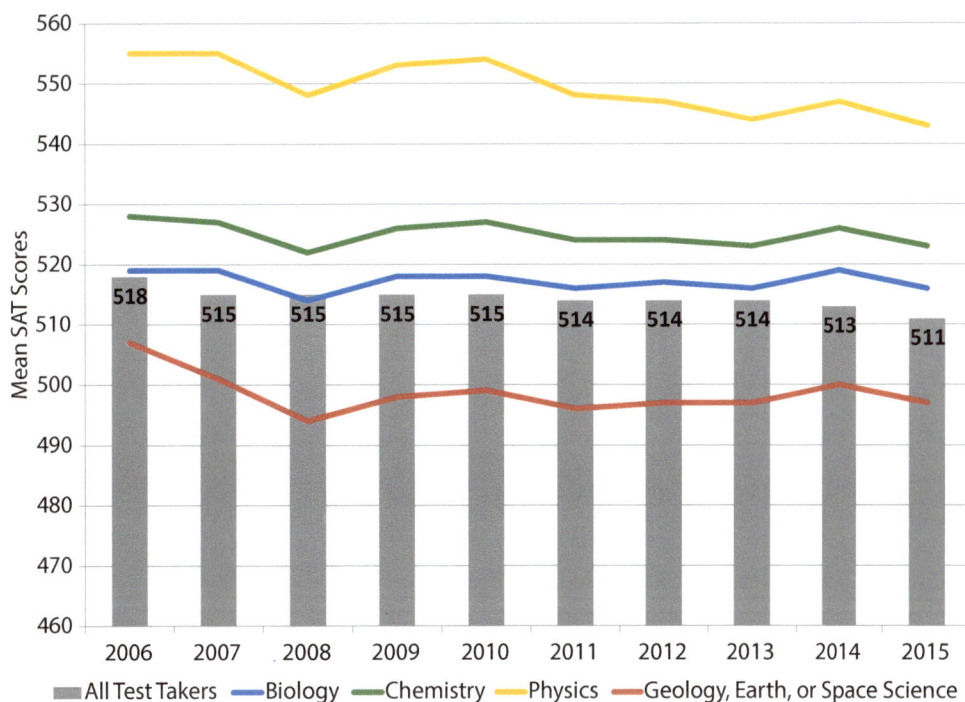

AGI Geoscience Workforce Program; Data derived from the College Board College-Bound Seniors, Total Group Report, 1996-2015

Figure 1.16: Mean Critical Reading SAT Scores for Test-Takers with Coursework in Science

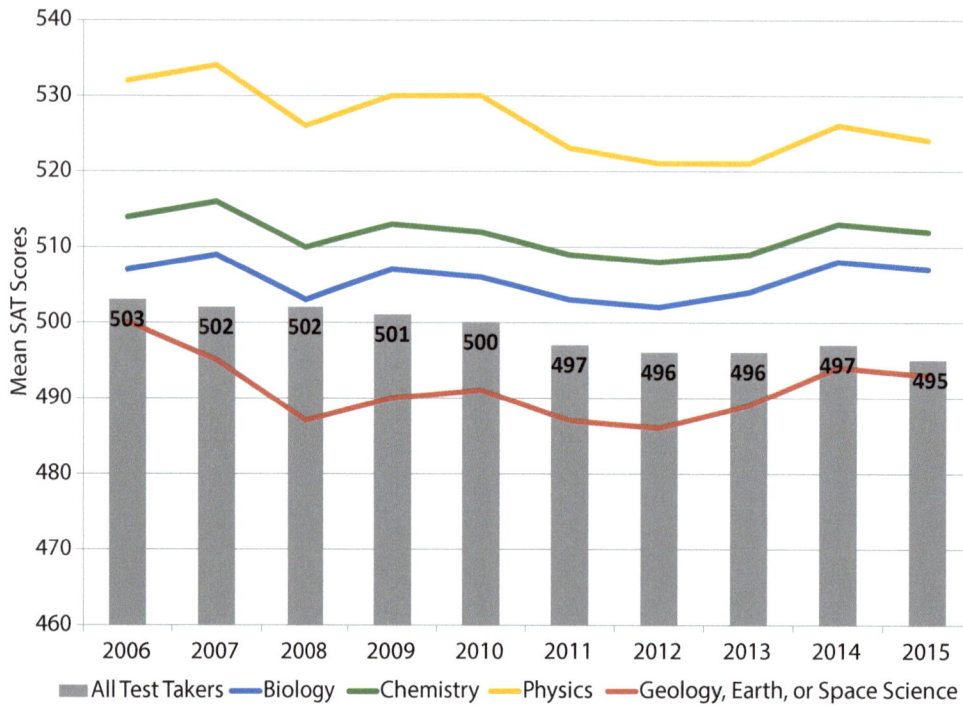

AGI Geoscience Workforce Program; Data derived from the College Board College-Bound Seniors, Total Group Report, 1996-2015

Figure 1.17: Mean Writing SAT Scores for Test-Takers with Coursework in Science

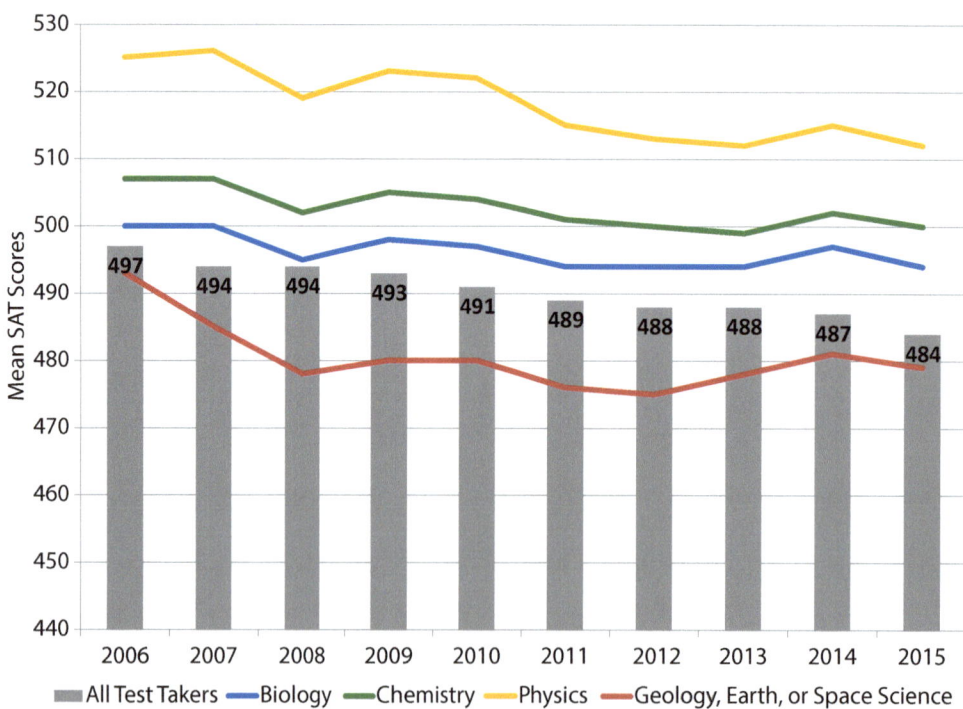

AGI Geoscience Workforce Program; Data derived from the College Board College-Bound Seniors, Total Group Report, 1996-2015

Figure 1.18: Intended Degree Level of College-Bound High School Seniors that took the SAT

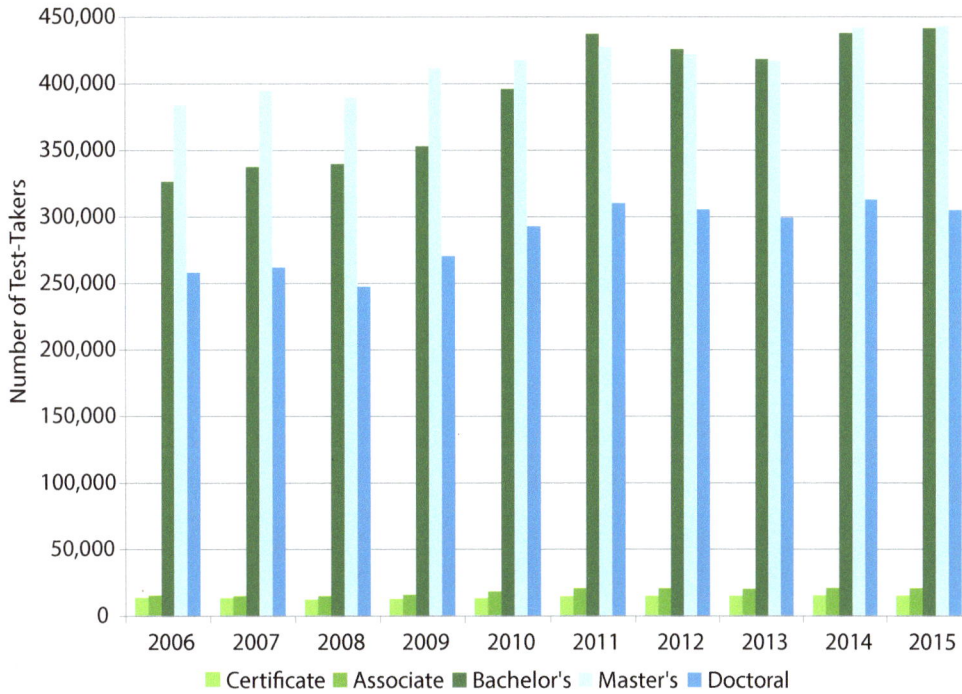

AGI Geoscience Workforce Program; Data derived from the College Board College-Bound Seniors, Total Group Report, 1996-2015

Figure 1.19: SAT Test-Takers Intending College Degrees in Physical Sciences or Interdisciplinary Studies

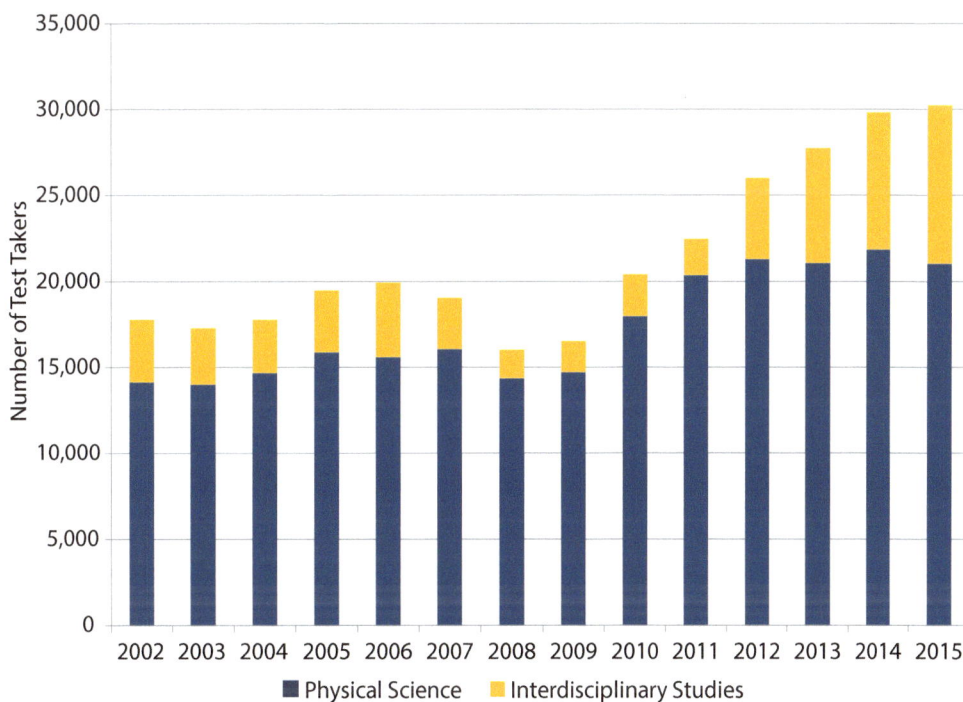

AGI Geoscience Workforce Program; Data derived from the College Board College-Bound Seniors, Total Group Report, 1996-2015

Figure 1.20: High School Students Meeting ACT College Readiness Benchmarks

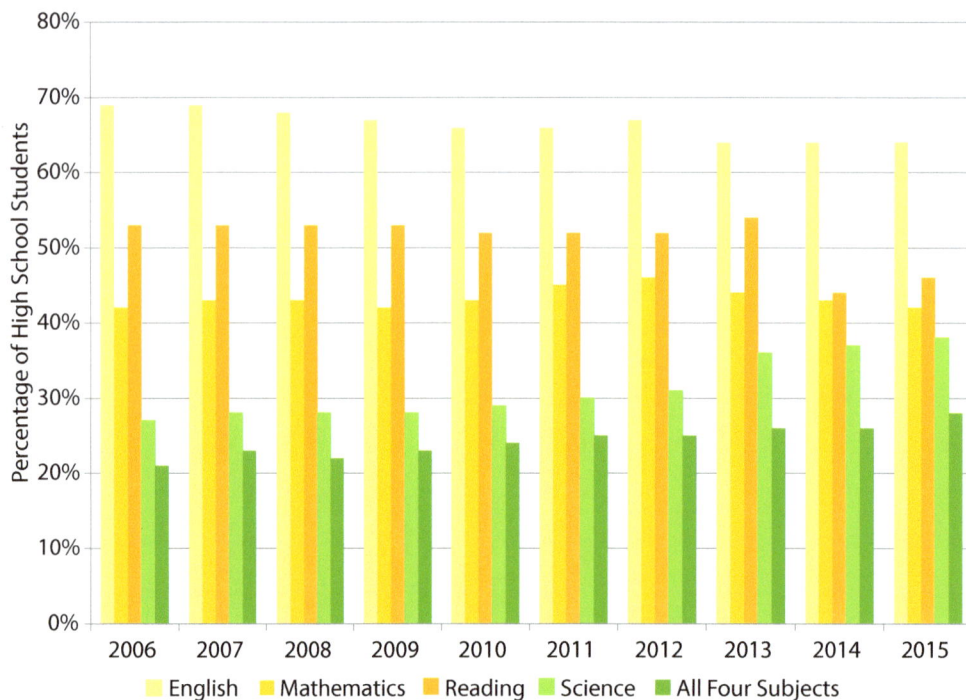

AGI Geoscience Workforce Program; Data derived from ACT National Profile Report, 2006-2015

Figure 1.21: Geoscience Postsecondary Graduates Who Took an Earth Science Course in High School

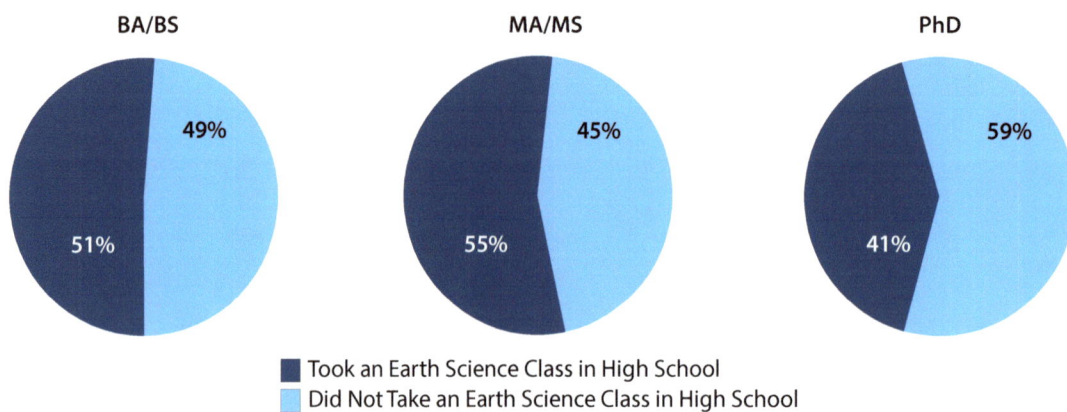

AGI Geoscience Workforce Program; Data derived from AGI's Geoscience Student Exit Survey Report 2015

Chapter 2: Trends in Two-Year College Geoscience Programs

Between 2010 and 2013, there was a decrease in the enrollments of students at two-year colleges, as well as little growth in the number of associate's degrees awarded. While the total number of students enrolled at two-year colleges did not increase, the percentage of women at these institutions remained above 55%, and Hispanic student enrollments continue to increase, reaching 22% in 2013. However, African American/Black and Native American/Native Indian enrollments have seen no growth between 2010 and 2013. These changes to enrollments and completions at two-year colleges have not affected the continued increase in the overall number of associate's degrees awarded in the geosciences, as well as the small gains in percentages of women and underrepresented minorities that earned associate's degrees in the geosciences.

Approximately 19% of two-year colleges in the United States offer a geoscience program or course. This may appear to be a decrease in programs since the 2014 report, but since the publication of the earlier edition of this report, the Directory of Geoscience Departments database has been extensively updated, which included the removal of duplicate listings. While AGI's database is extensive, it may not contain all the geoscience programs in existence at two-year colleges due to periodic changes in programs, course offerings, and faculty. There is an identifiable relationship between states with major geoscience industries and the number of 2-year institutions with a geoscience presence, such as Texas, California, Washington, Illinois, New York, and Arizona.

Most geoscience activities within two-year colleges only have one or two faculty members assigned to teach a geoscience course, and these faculty members tend to be within a natural sciences or physical sciences division. Two-year colleges tend to have younger faculty and a higher percentages of female faculty in tenure and tenure-track positions than at four-year institutions.

Geoscience faculty teaching at two-year colleges have indicated that there is a growing interest in the geosciences among their students, particularly in states with strong geoscience industries, and they tend to encourage their students to transfer to four-year institutions to complete their geoscience education. Therefore, this student population is an ideal target for recruitment of geoscience majors at four-year institutions. Over the past three years, there has been a steady increase in the percentages of bachelor's, master's, and doctoral graduates that indicated spending at least one semester at a two-year college during their postsecondary education. Among these graduates, there have also been increases in the percentages of these recent students taking an earth science course while at the two-year colleges. Further development of relationships between two-year colleges and four-year institutions will increase the recruitment of students into geoscience majors.

With growing evidence for more active recruitment of geoscience majors from two-year college transfers, AGI is currently collaborating with two-year college faculty to investigate the factors that assist in effective transfer of two-year college students, as well as the challenges these students have faced after entering a geoscience program at a four-year institution. While family and friends were seen as helpful to some students with their transition into four-year institutions, the majority of two-year college transfers considered their personal motivation to a bachelor's degree and the transferred coursework as the most impactful for a successful transfer and completion. This research is ongoing as more information related to the transfers of two-year college students is released.

National Benchmarks

Changes in general population dynamics highlight the ever growing talent pool of college-age underrepresented minority groups, particularly Hispanics, and women. By 2060, the Hispanic college-age population of is projected to grow to within 12% of the Caucasian college-age population (Figure 2.1). Currently 49% of the U.S. college-age population is female, and the majority of all degrees are awarded to women (Figure 2.3). This is particularly true for two-year colleges with the increasing percentages of associate's degrees awarded to underrepresented minorities and the consistently high percentage of women earning associate's degrees (Figures 2.3 and 2.4). In 2013, the percentage of associates degrees awarded to Hispanic students was higher that the percentage of African American students for the first time since

1977 (Figure 2.4). Figures 2.2 and 2.3 present projected data based on the most recent data reported for 2014.

From 2012-2015, there appears to have been very little to no growth in the number of postsecondary degrees at all degree levels (Figure 2.2). This raises the question of what is driving this stagnation of the number of degrees awarded. Have the U.S. postsecondary institutions approached capacity? Is the rapidly increasing cost of tuition affecting the number of completions at these institutions? While the numbers increase beyond 2015, these values are projections based on the most recent data from 2013. Therefore, whatever factors are affecting the increase in total completions could continue into the future or they could be artifacts of the economy during this time period.

Figure 2.1: Race/Ethnicity of U.S. College-Age Population, 2014-2060

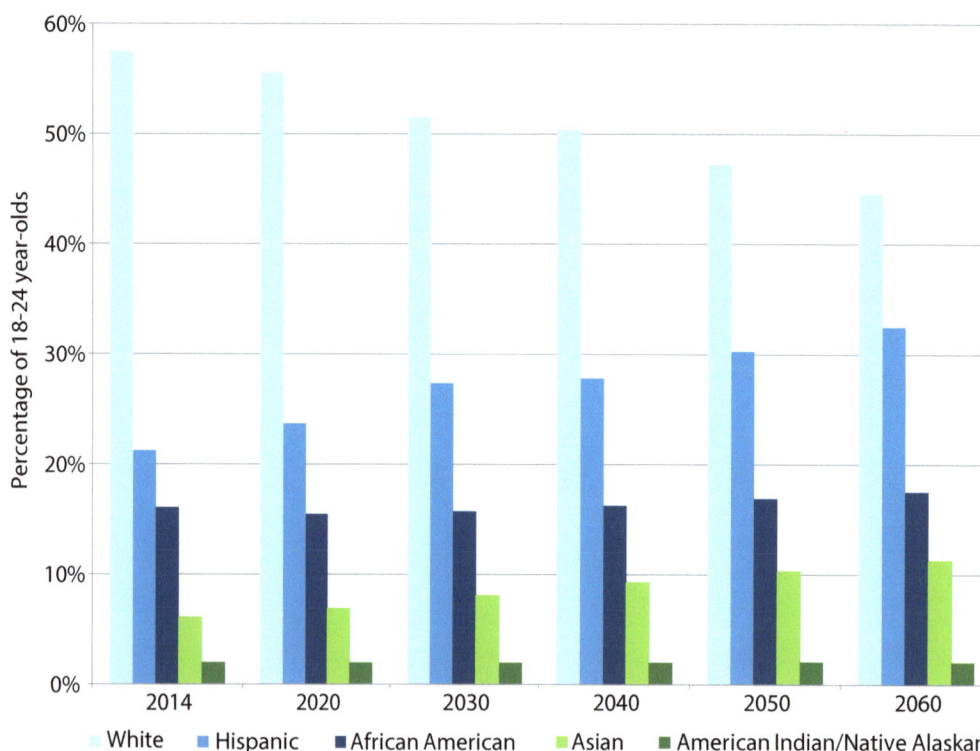

Figure 2.2: Degrees Granted from U.S. Postsecondary Institutions

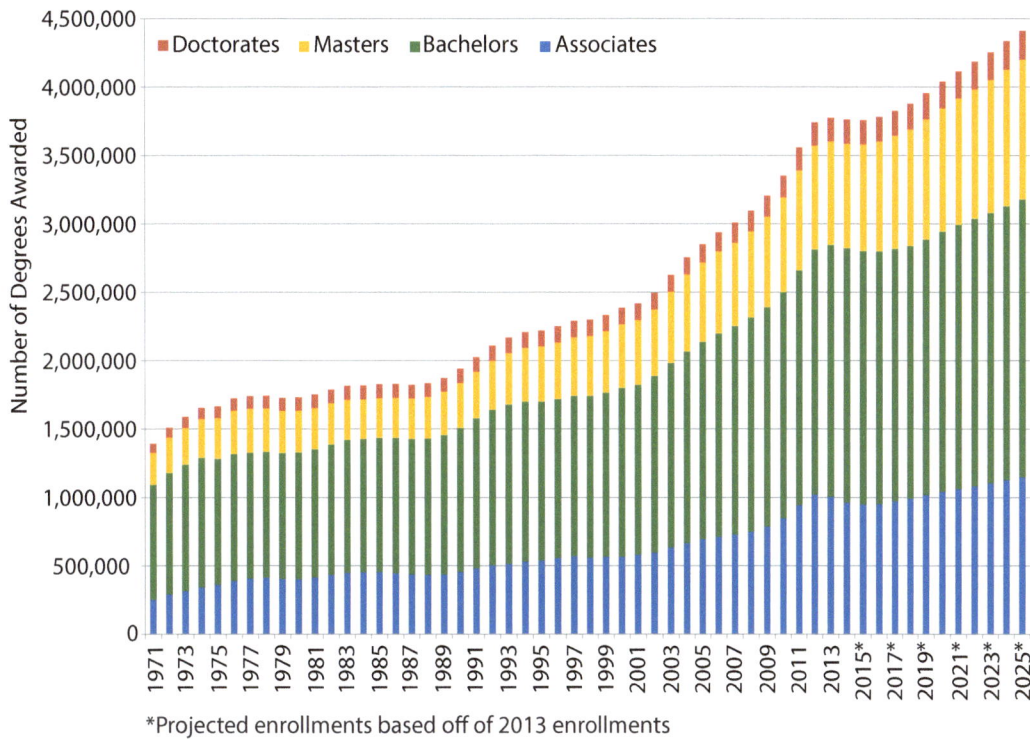

*Projected enrollments based off of 2013 enrollments

AGI Geoscience Workforce Program; Data derived from NCES Digest of Education Statistics, 2014

Figure 2.3: Percentage of Degrees Granted to Women by Degree Level, All Majors

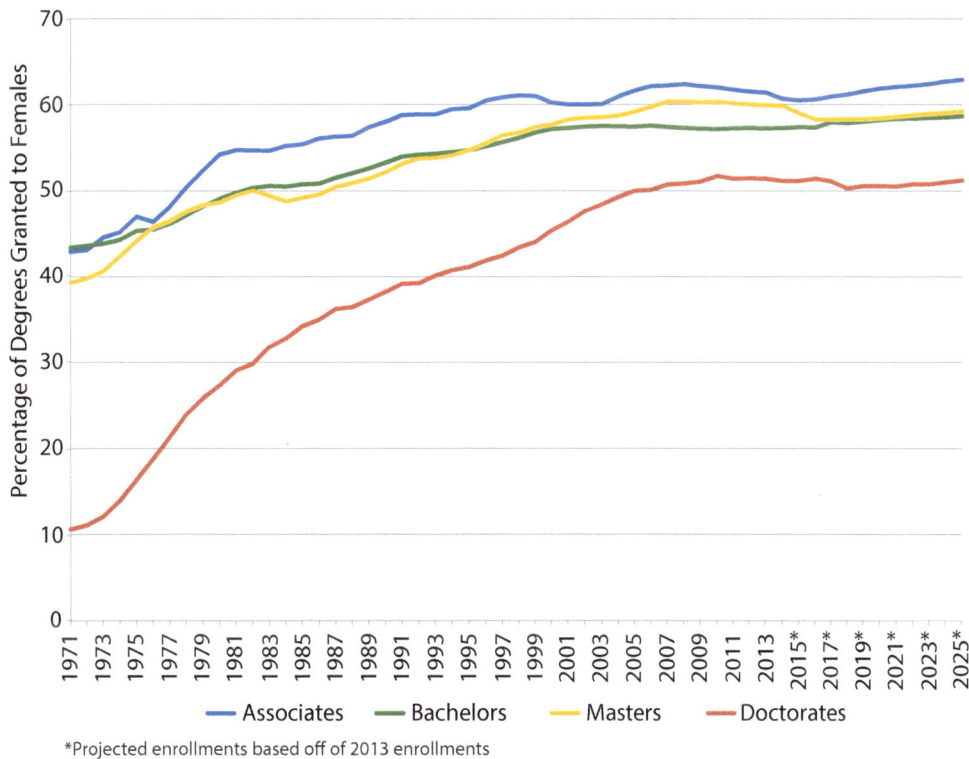

*Projected enrollments based off of 2013 enrollments

AGI Geoscience Workforce Program; Data derived from NCES Digest of Education Statistics, 2014

Figure 2.4: Percentage of Associate's Degrees Awarded to Underrepresented Minorities, All Degree Fields

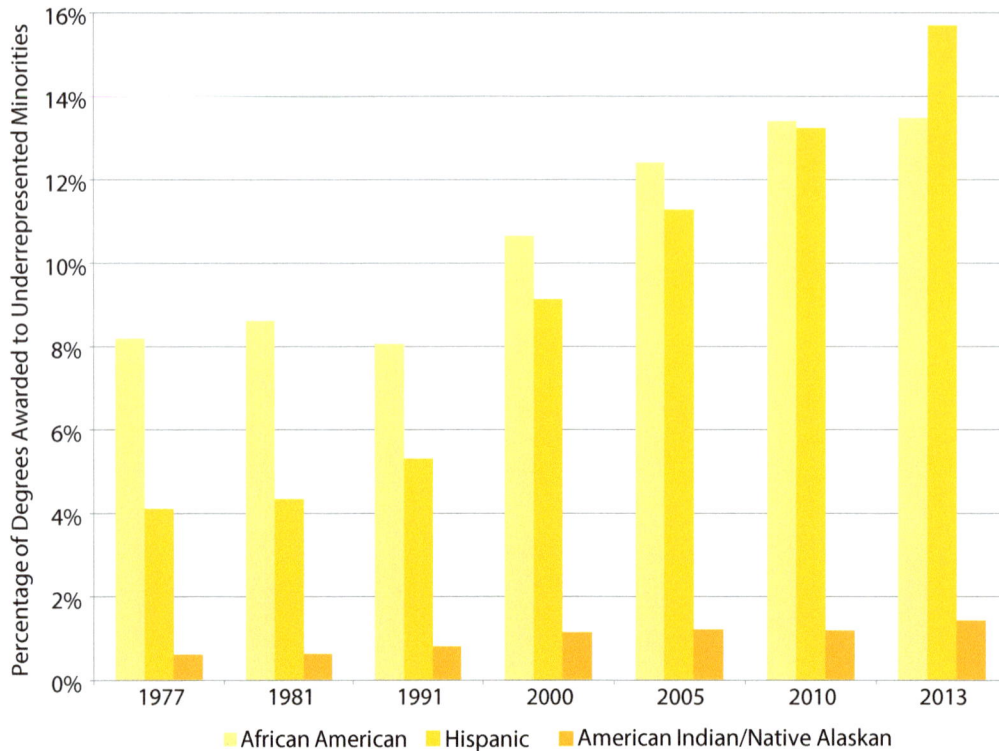

AGI Geoscience Workforce Program; Data derived from NCES Digest of Education Statistics, 2014

Two-Year Enrollments

After 2011, fall enrollments at two-year colleges in the United States dropped from approximately 7.5 million students in 2011 to approximately 7 million students in 2013, which coincides with the lack of growth in completions during the same time period (Figure 2.5). Something changed over the past few years affecting both enrollments and degrees awarded at two-year colleges in the U.S. Female enrollments remain steady at just under 60% (Figure 2.6). Enrollment percentages of Hispanic students continue to rise at two-year colleges, but the enrollments percentages of African Americans and Native Americans have not changed sing 2010 (Figure 2.7). With the rising percentages of Hispanic enrollments at two-year enrollments and the steadily high enrollments percentage of women, more focus needs to turn to recruitment of these students into geoscience majors and career fields. However consideration is needed for those factors that have been affecting enrollments and completions at two-year colleges, particularly if this trend continues in the future.

Figure 2.5: Fall Enrollments at U.S. Two-Year Colleges

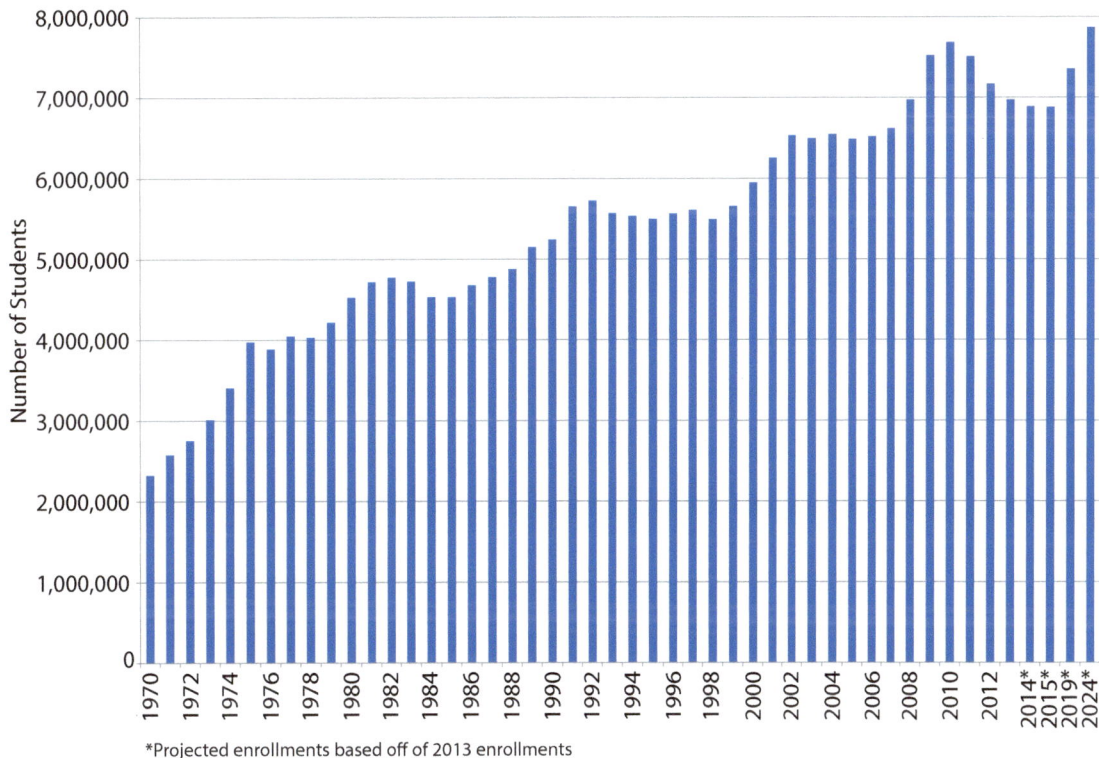

*Projected enrollments based off of 2013 enrollments

AGI Geoscience Workforce Program; Data derived from NCES Digest of Education Statistics, 2014

Figure 2.6: Participation of Women in Two-Year Colleges

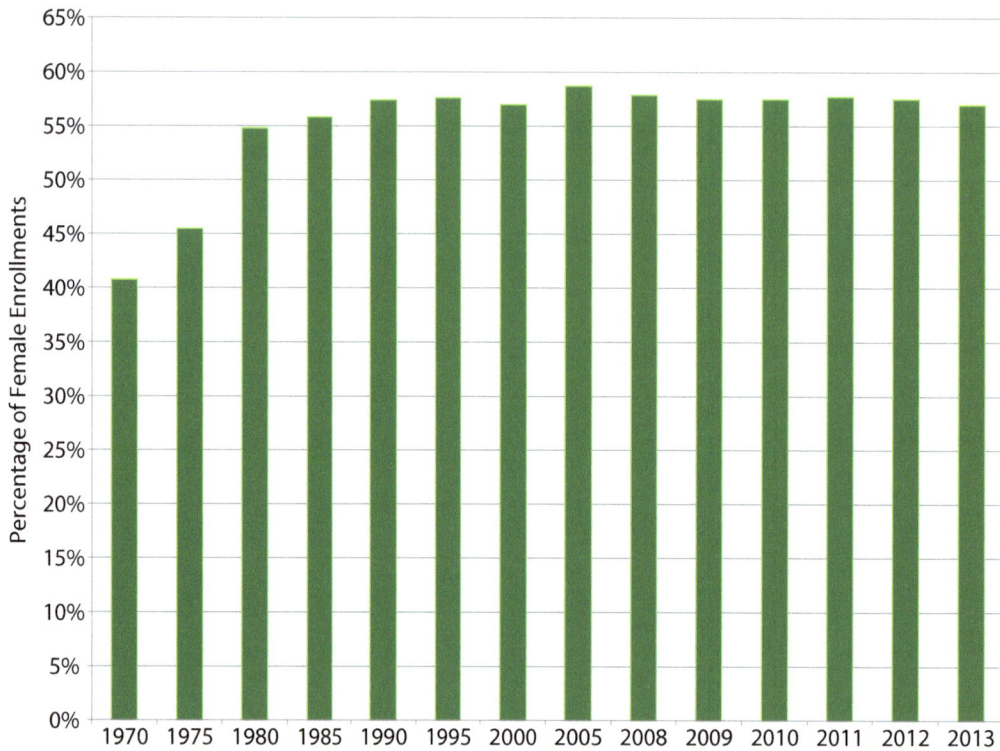

AGI Geoscience Workforce Program; Data derived from NCES Digest of Education Statistics, 2014

Figure 2.7: Underrepresented Minority Enrollments at Two-Year Colleges

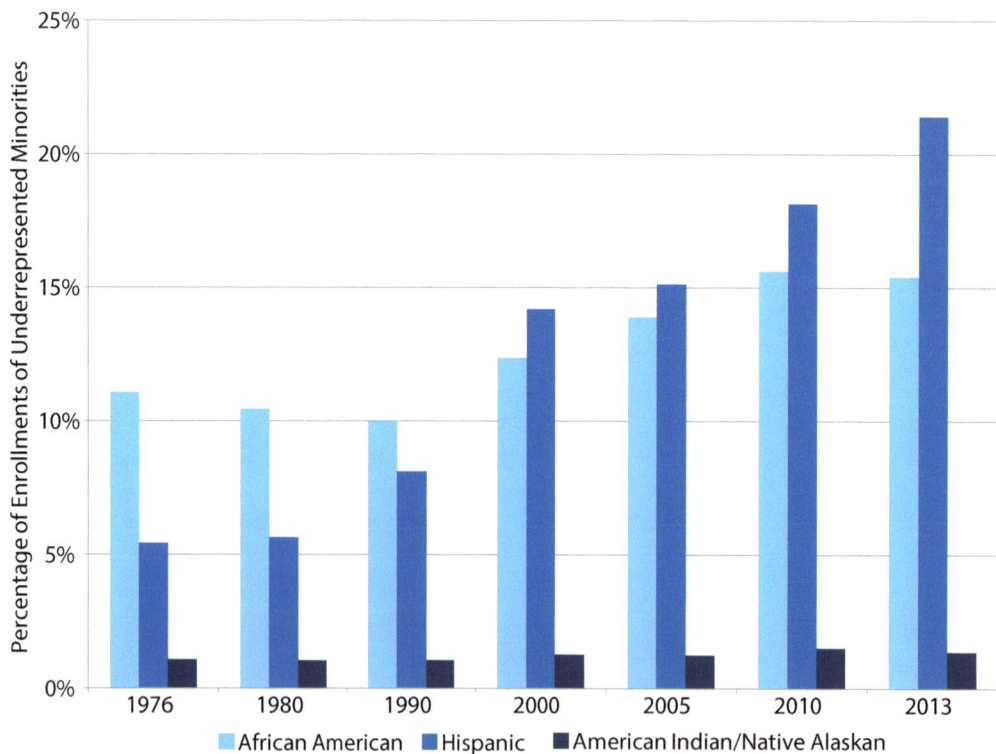

AGI Geoscience Workforce Program; Data derived from NCES Digest of Education Statistics, 2014

Geoscience Departments and Faculty

There are currently 1,700 two-year colleges in the United States, according to the Department of Education, and 329 of these schools have a geoscience program or course available for students (Figure 2.8). The total number of geoscience programs in the U.S. is lower than the number presented in the 2014 report because of some double counting of programs. California and Texas continue to have the highest number of geoscience programs or courses available at two-year colleges, but Washington has the highest percentage of geoscience programs within all of the two-year colleges in the state (Table 2.1).

However, most of these schools that provide geoscience courses have very few faculty to teach these classes. Out of the 329 two-year colleges with geosciences, 79% have less than 5 faculty members that teach geoscience coursework and 38% have only one faculty member to represent the geosciences (Figure 2.9). Approximately 39% of the geoscience faculty teaching at two-year colleges is working part-time at these institutions (Figure 2.10). Thirty percent of two-year college geoscience faculty is female, compared to the 19% of four-year university geoscience faculty (Figure 2.11 and 3.4). Both are increases on the percentages presented in the 2014 report. The female faculty population in two-year colleges and four-year universities has been slowly increasing since 2008. The increased percentage of female faculty members at two-year colleges can be seen across Associate Professors, Instructors/Lecturers, Adjunct/Visiting Professors, and Emeritus faculty.

Figure 2.8: Number of Geoscience Departments/Programs at Two-Year Colleges by State

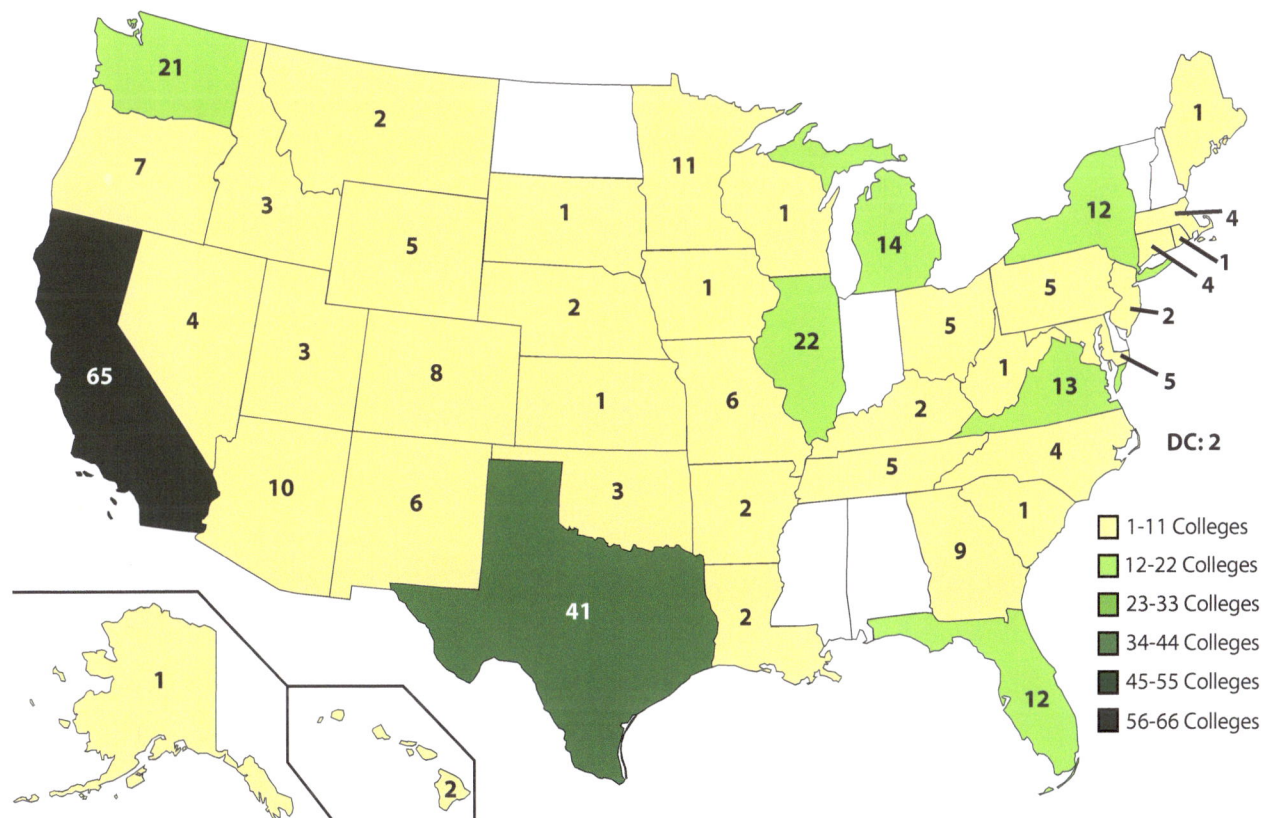

AGI Geoscience Workforce Program; Data derived from AGI's Directory of Geoscience Departments database

Table 2.1: Percentage of Two-Year Colleges with Geoscience Programs for Selected States

State	2-Year Colleges with Geoscience Programs	2-Year Colleges in the State	Percentages of 2-Year Colleges in the State with Geoscience Programs
California	65	193	34%
Texas	41	125	33%
Illinois	22	65	34%
Washington	21	43	49%
Michigan	14	32	44%
Virginia	13	47	28%
Florida	12	88	14%
New York	12	75	16%
Minnesota	11	33	33%
Arizona	10	35	29%

AGI Geoscience Workforce Program; Data derived from AGI's Directory of Geoscience Departments database and Carnegie Classification of Institutions of Higher Education

Figure 2.9: Number of Faculty per Geoscience Department/Program at Two-Year Colleges

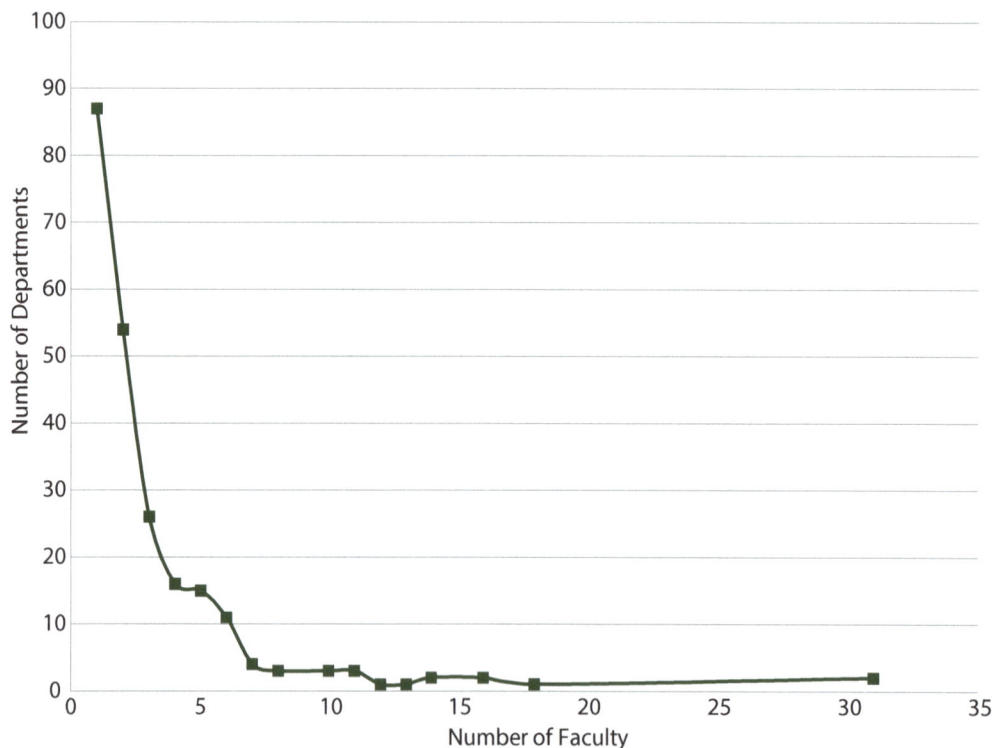

AGI Geoscience Workforce Program; Data derived from AGI's Directory of Geoscience Departments database

Figure 2.10: Age Demographics of Two-Year College Geoscience Faculty

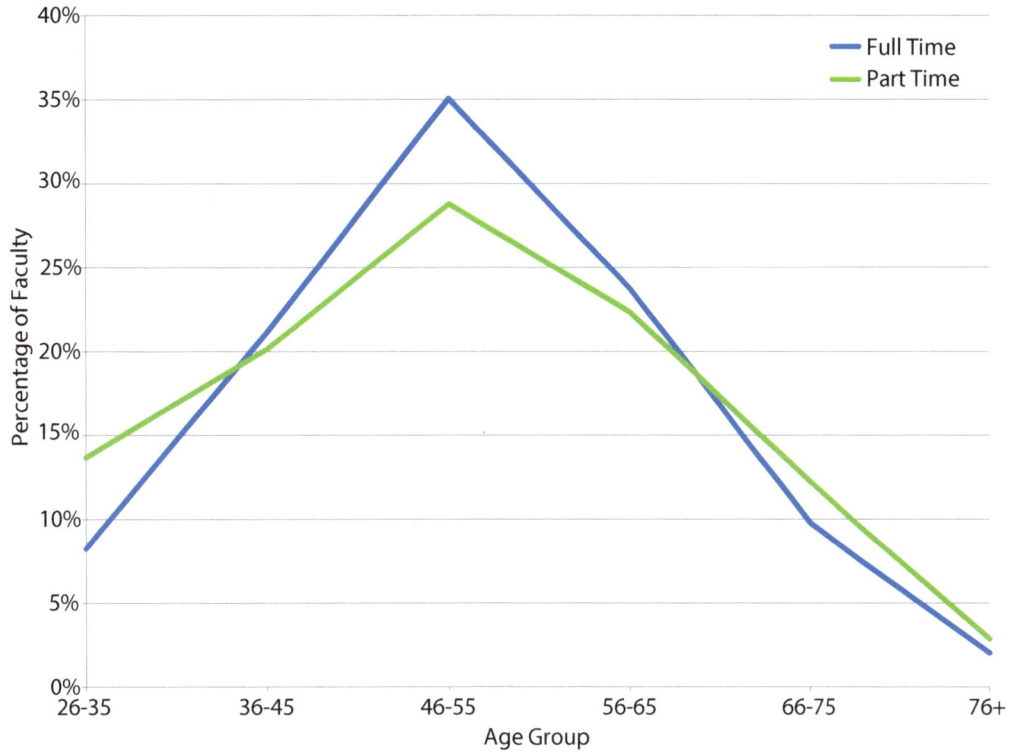

AGI Geoscience Workforce Program; Data derived from AGI's Directory of Geoscience Departments

Figure 2.11: Percentage of Two-Year Geoscience Faculty Positions Held by Women

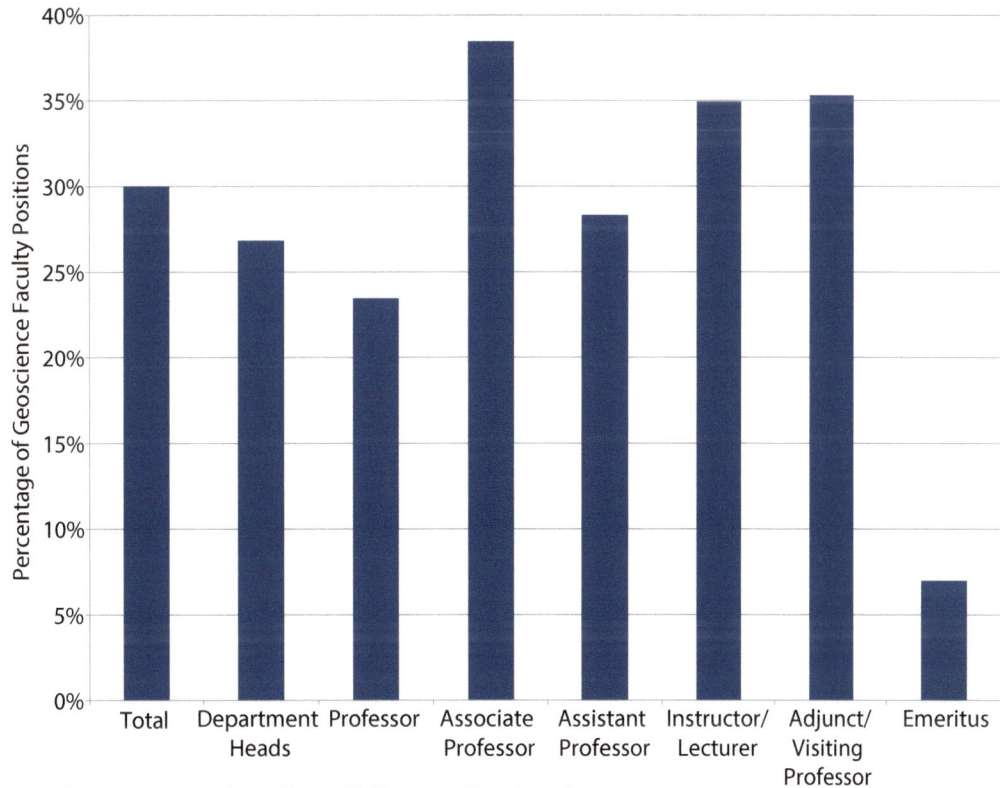

AGI Geoscience Workforce Program; Data derived from AGI's Directory of Geoscience Departments

Geoscience Students

Many of the students that attend a two-year college transfer to four-year universities with or without completing an associate's degree, particularly in the sciences. According to the National Science Foundation, just over half of bachelor's graduates, 35% of master's graduates, and 20% of doctoral graduates with a degree in the physical and related sciences attended a two-year college (Figure 2.12). In the geosciences, 33% of bachelor's graduates, 23% of master's graduates, and 16% of doctoral graduates spent at least one semester at a two-year college (Figure 2.13). There percentages

have grown each year since 2013. There has also been growth in the percentage of geoscience graduates that took a geoscience course at a two-year college from 14% in 2013 to 19% in 2015 of bachelor's graduates and from 5% in 2013 to 12% in 2015 of master's graduates. These figures also reinforce the need to increase recruitment of geoscience majors among two-year college students. These students are developing an interest in the subject through the geoscience courses at two-year colleges and a growing number of these students are carrying that interest into four-year universities.

Figure 2.12: Four-Year University Graduates by Degree Field Who Attended a Two-Year College, 2013

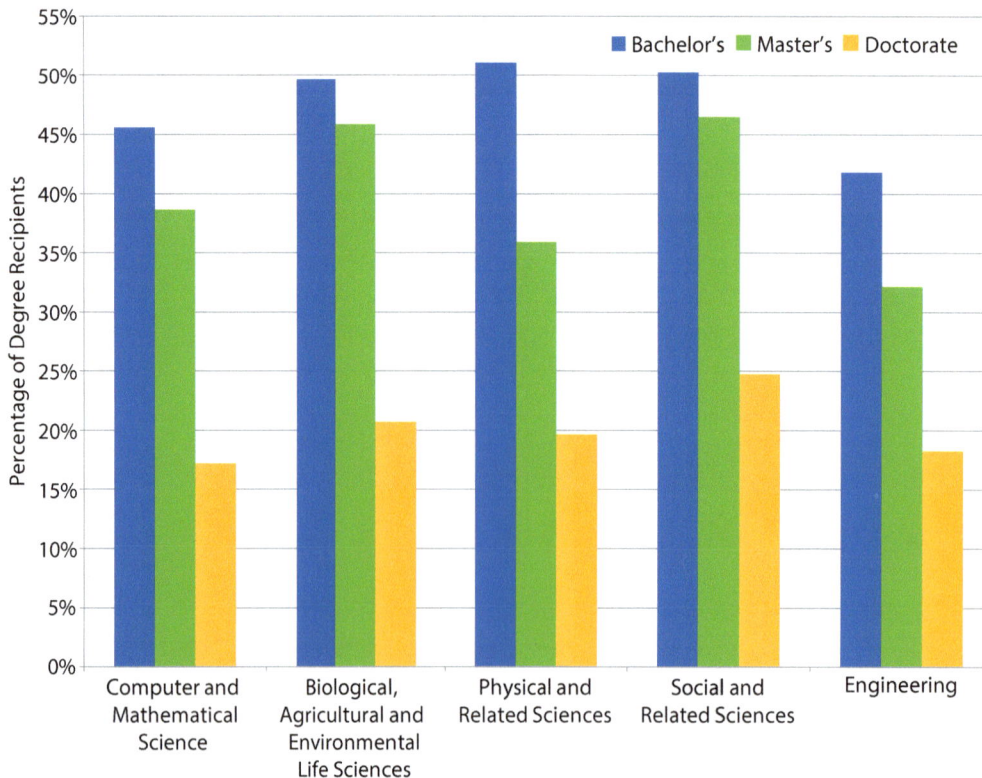

AGI Geoscience Workforce Program; data derived from NSF's SESTAT 2013 Public Dataset

Figure 2.13: Geoscience Graduates with at Least One Semester at a Two-Year College, 2015

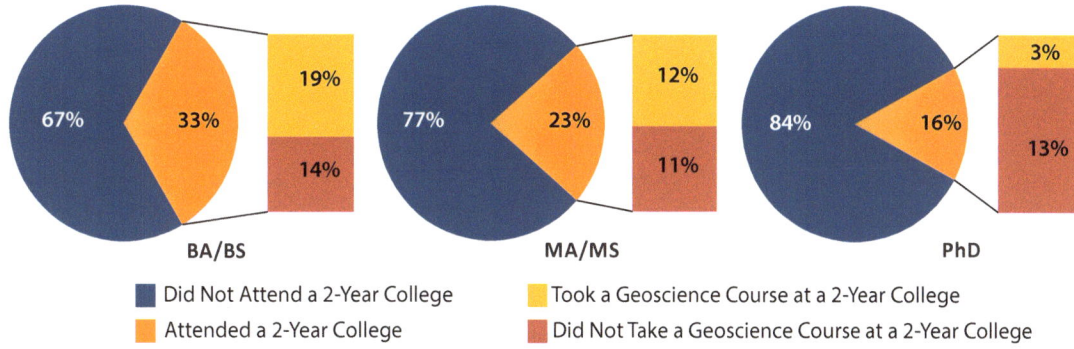

BA/BS   MA/MS   PhD

■ Did Not Attend a 2-Year College   ■ Took a Geoscience Course at a 2-Year College
■ Attended a 2-Year College   ■ Did Not Take a Geoscience Course at a 2-Year College

AGI Geoscience Workforce Program; Data derived from AGI's Geoscience Student Exit Survey

Geoscience Associate's Degrees

While enrollments at two-year colleges have been decreasing and completions have seen no growth over the past few years, the number of awarded geoscience associate degrees has increased by 60% from 239 in 2012 to 382 in 2014 (Figure 2.14). Of the 382 geoscience associate's degrees awarded in 2014, 38% were awarded to women and 20% were awarded to underrepresented minorities—a 2% increase for both groups since 2012 (Figures 2.15 and 2.16).

All three figures were derived from data provided by the Department of Education's Integrated Postsecondary Education Database System (IPEDS). When looking at these data, it is important to remember that the information is self-reported by each institution's main administrative office. The information represents the main offices' definitions of geoscience and other fields. This may help explain some of the change and variance in the number of geoscience associate's degrees awarded each year. This information would be more reliable if reported directly from the geoscience programs within the two-year colleges, but with many of these programs consisting of only one or two faculty members, this data collection would be difficult to initiate and sustain.

Figure 2.14: Geoscience Associate's Degrees Awarded Annually

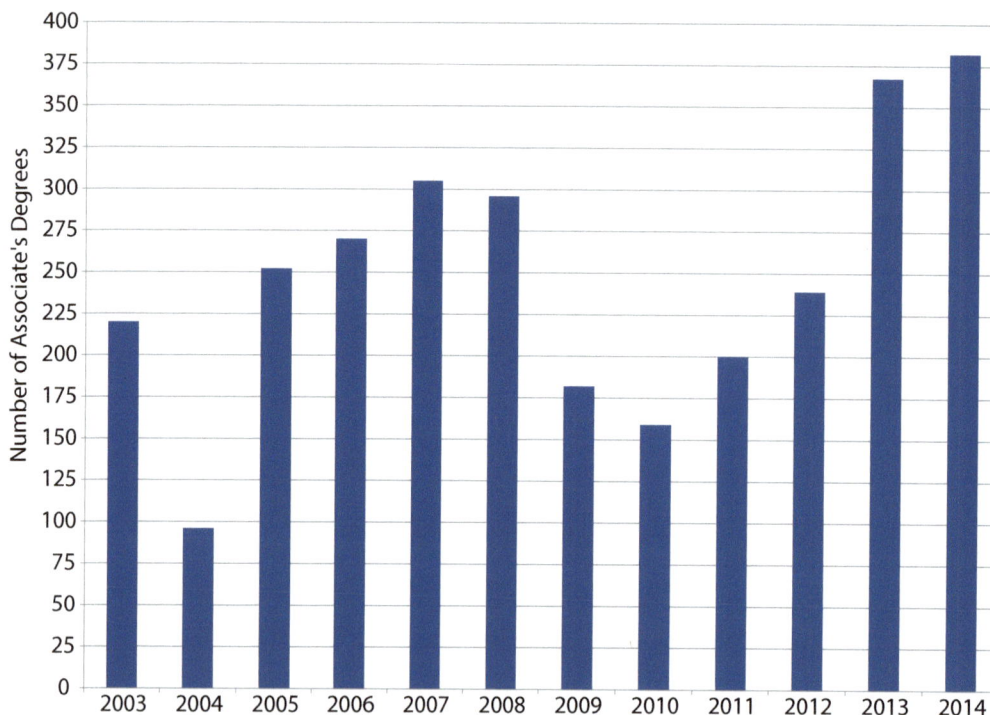

AGI Geoscience Workforce Program; Data derived from IPEDS

Figure 2.15: Percentage of Associate's Degrees Awarded to Women by Discipline

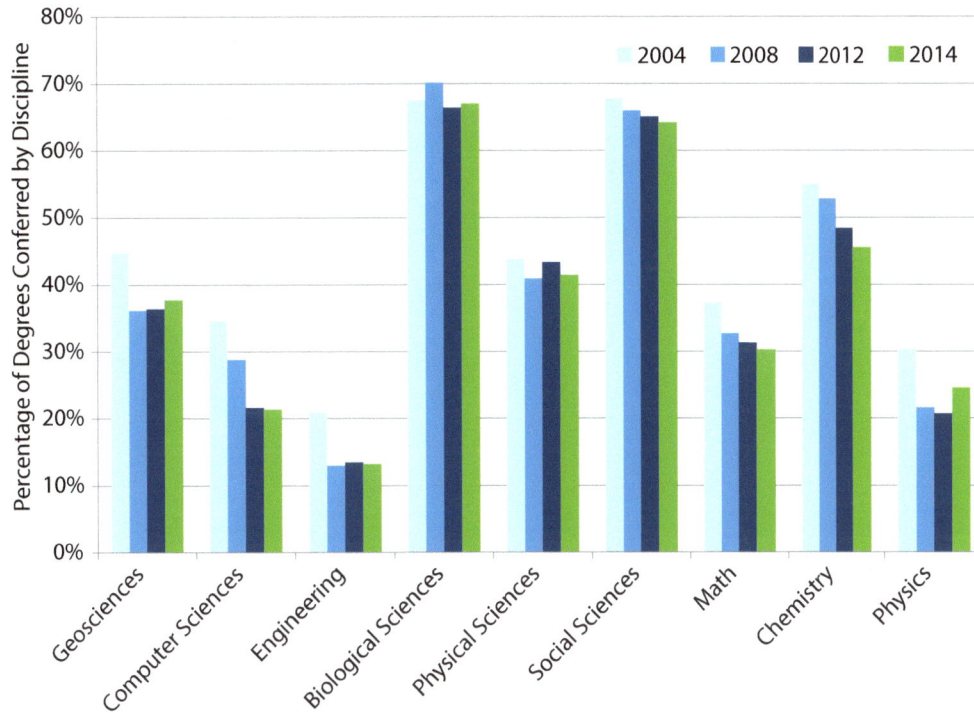

AGI Geoscience Workforce Program; Data derived from IPEDS

Figure 2.16: Percentage of Associate's Degrees Awarded to Underrepresented Minorities by Discipline

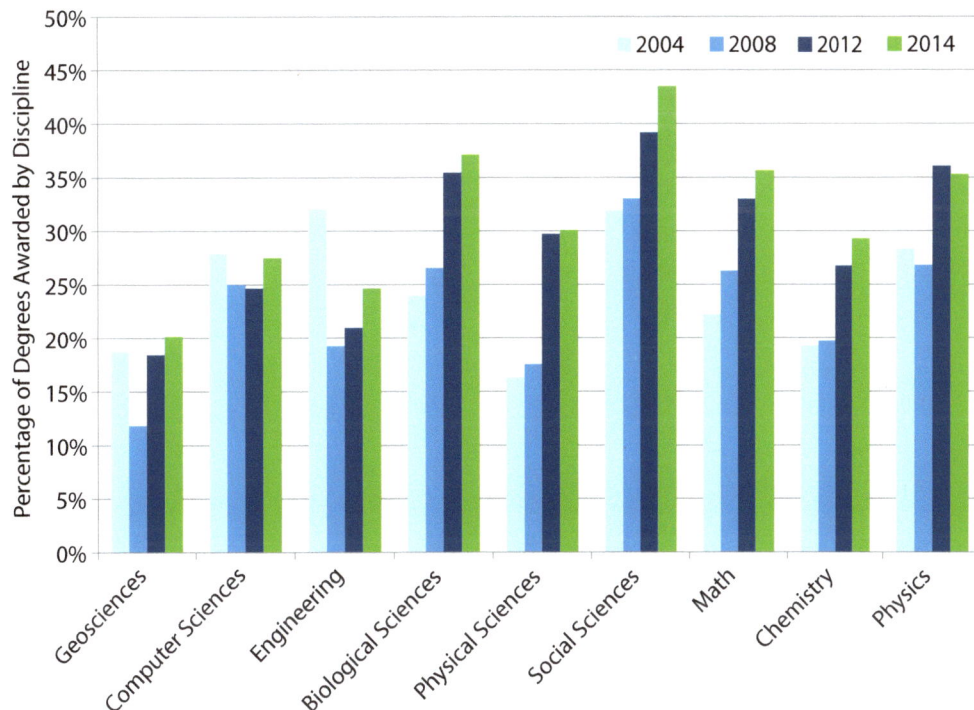

AGI Geoscience Workforce Program; Data derived from IPEDS

Chapter 3: Trends in Four-Year Institution Geoscience Programs

A geoscience degree from a four-year institution is essential for developing the necessary knowledge and technical skills to be successful in the workforce as a geoscientist. The master's degree historically has been considered the degree of employment for private sector jobs in the geosciences, but recently bachelor recipients are starting to be hired more frequently. Therefore measuring the health of geoscience departments at four-year institutions is essential for tracking the future of the geoscience workforce.

Geoscience faculty at four-year institutions tend to enter tenure-track positions 2-4 years after completing their doctorate, and typically takes approximately 20 years from completion of the doctorate to reach full professor. Since 2013, the number of emeritus faculty has decreased from just over 1,100 faculty members to 1,000 faculty members. Approximately 67% of tenure-track geoscience faculty are above federal retirement age, and 19% of all geoscience faculty are listed as emeritus. The percentages of females in geoscience faculty and researcher positions continue to slowly increase to 19% of the entire geoscience workforce at four-year institutions. There has been a steady biennial increase of 1-2% of women in faculty positions since 2008 at the various ranks in the academic workforce, and this trend is expected to continue due to the relatively high percentage of women currently in assistant professor positions.

Over the past few years, four- year institutions have seen little to no growth in enrollments and degrees granted, which hints at some sort of economic or national issue that may be affecting the participation in postsecondary education. The same trend has been seen at two-year colleges as well. However, in the geosciences, undergraduate enrollments continue to rise and awarded degrees to bachelors and masters graduates have increased as well. Geoscience departments did not seem to be impacted by the complication affecting overall enrollments and completions in postsecondary education. Instead, historically, geoscience enrollments tend to follow the economic trends of the geosciences, and while the oil and gas industry has been experiencing an economic downturn over the past couple of years, the environmental and mining industry have grown in their recruitment of recent graduates into employment. Some of the growth in enrollments can be attributed to an increase in the development of online degree programs for the geosciences at four-year institutions. It is not yet clear how these programs will affect the supply of graduates to the workforce, and research is needed to consider the impact of these programs.

Enrollments in graduate programs have been hovering around 10,000 students for the past thirty years indicating that these programs have been at capacity for some time now.

Female enrollments in the geosciences appear to have been on the decline since 2003 even while total enrollments have increased and the participation of women at four-year institutions is consistently around 57%. Even though there appears to be a decrease in the percentage of women in geoscience degree programs, the actual number of females enrollments has been increasing, just not as fast as male enrollments. Students from underrepresented minorities are harder to track through enrollments; however data collection efforts by AGI and NSF provide a better understanding of the graduation rates of minorities with geoscience degrees. The bachelor's degree completions of underrepresented minorities in the geosciences have been slowly increasing over the past few years reaching approximately 11% in 2015, but this increase can be almost solely attributed to an increase in the participation of Hispanic students in geoscience departments. The geosciences still tend to have the lowest participation rates from underrepresented minorities compared to all science disciplines according to NSF. While there are currently many efforts aimed at increasing the underrepresented population of students on the geosciences, there may also need to be more focus on encouraging women to major in the geosciences. An alternate consideration for the recruitment of diverse populations into the geosciences may be through a concerted effort of recruitment of first generation college students. A large majority of geoscience graduates had at least one parent or guardian with a postsecondary degree, while 10-13% of graduates were first generation college students. First generation college students also tend to be from lower socioeconomic categories and could benefit from the help and funding provided for recruitment of new majors.

Nearly every geoscience student participates in a field experience, whether it is field camp, a field course, or a field component during a trip or outing. Field camp attendance reached a peak in number of students participating in 2014 with a decrease in participation in 2015. Field camps seem to be at capacity, but it is still strongly encouraged or required for participation. Field camp access may become a limiting factor for some graduate's employability. AGI has noted field camps sponsored by departments with open and closed enrollment in the Directory of Geoscience Departments in order to help students locate and participate in field camp.

Field experiences and research experiences are highly encouraged, or required, activities for geoscience students at all degree levels, but student internships do not seem to be as emphasized within geosciences programs, as seen by the low participation rates, particularly at the bachelor's and doctoral degree levels. Approximately 60% of master's graduates participated in at least one internship, and all students that participate in an internship find the activity very important for their academic and professional development. Discussions with industry representatives have highlighted the need for more students to gain professional skills through activities, such as internships, before entering the workforce because that experience is critical to be successful in the geoscience workforce.

Much of the data presented in this report on geoscience students at four-year institutions originated from AGI's Geoscience Student Exit Survey. This study was developed to understand the preparation of students for entering the workforce and their immediate plans after graduation. The annual reports, with more in depth analysis of the results, can be found at http://www.americangeosciences. org/workforce/reports.

Geoscience Departments

According to AGI's Directory of Geoscience Departments, there are currently 638 geoscience departments at four-year universities currently (Figure 3.1). Since the 2014 edition of this report, a concerted effort was conducted to update the geoscience department listings in AGI's Directory of Geoscience Departments database. Changes in the number of departments since 2014 is partially from quality control efforts which included the removal of ambiguous programs and merged programs, as well as some closures of departments.

Changes in faculty appointments and student enrollments have impacted the student to faculty ratio compared to the 2014 edition of the report. Therefore, we reported in table form the top 10 highest student to faculty ratios in the U.S. instead of the former map showing these ratios in all states

(Table 3.1). In the U.S., the average number of geoscience tenure-track faculty within a geoscience department is approximately 10 faculty members, and the average enrollment within geoscience departments is 33 students. Texas leads the nation in the number of geoscience undergraduate and graduate students enrolled within geoscience departments, followed closely by Colorado and Oklahoma for undergraduates and California and Colorado for graduate enrollments (Tables 3.2 and 3.3).

The data provided to AGI's Directory of Geoscience Departments is self-reported by representatives within each department on a near-annual basis. Over the past two years, AGI has actually increased the number of departments that are regularly updating their listings on a yearly basis, which has strengthened the reliability of the data.

Figure 3.1: Number of Geoscience Departments by State

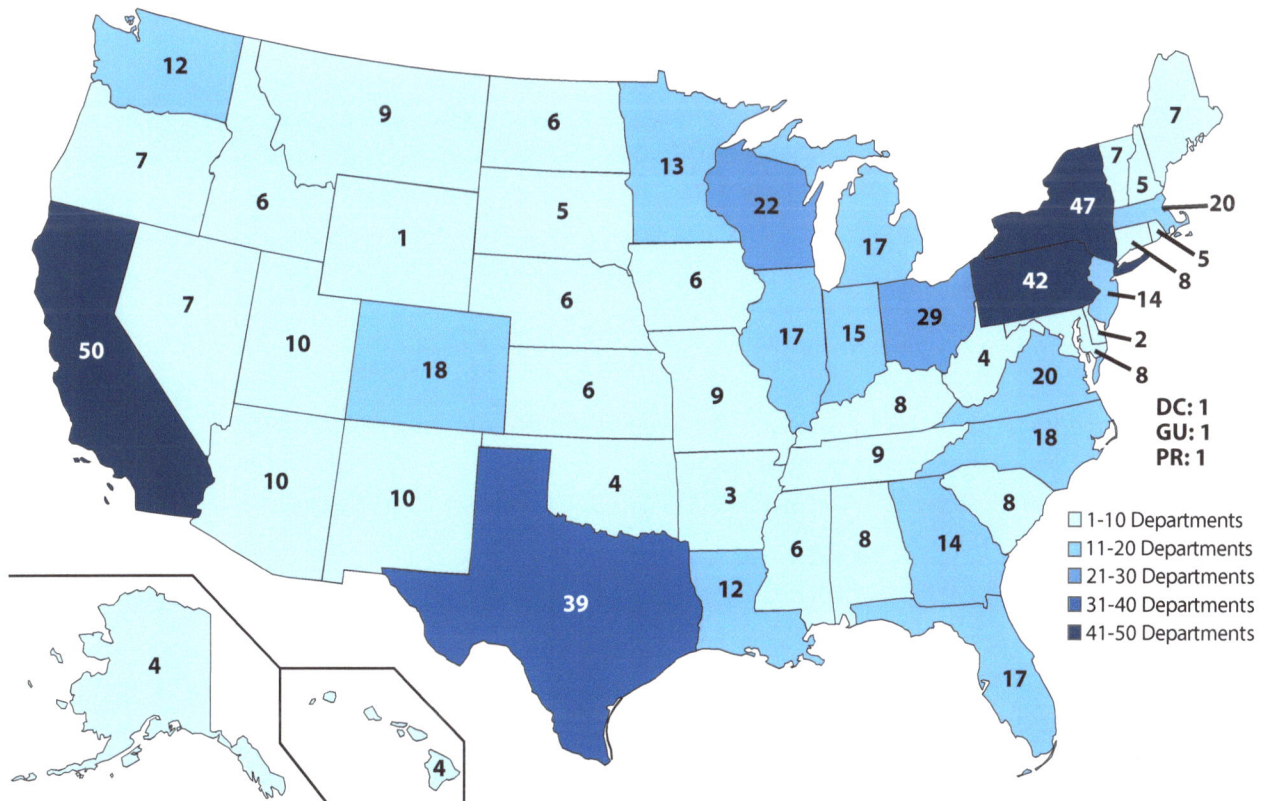

AGI Geoscience Workforce Program; Data derived from AGI's Directory of Geoscience Departments database

Table 3.1: Universities with the Highest Student per Faculty Member Ratios, 2015

State	Number of Students per Tenure Track Faculty Member
Texas	24
New York	18
Oklahoma	17
Pennsylvania	17
Mississippi	13
North Carolina	13
Illinois	11
Colorado	10
Ohio	9
Minnesota	8

AGI Geoscience Workforce Program; Data derived from AGI's Directory of Geoscience Departments database

Table 3.2: Percentage of All U.S. Geoscience Undergraduate Students Enrolled in 2014-2015

State	Percentage of All Undergraduate Geoscience Students
Texas	7.7%
Colorado	6.8%
Oklahoma	6.0%
California	5.7%
Pennsylvania	5.3%
New York	5.3%

AGI Geoscience Workforce Program; Data derived from AGI's Directory of Geoscience Departments database

Table 3.3: Percentage of All U.S. Geoscience Graduate Students Enrolled in 2014-2015

State	Percentage of All Graduate Geoscience Students
Texas	11.8%
California	8.5%
Colorado	8.4%
Oklahoma	4.6%
Pennsylvania	4.4%
Illinois	4.0%

AGI Geoscience Workforce Program; Data derived from AGI's Directory of Geoscience Departments database

Status of the Geoscience Workforce 2016

Geoscience Faculty

In 2015, there were 10,048 geoscience faculty and researchers employed in U.S. geoscience departments at four-year universities, compared to 10,265 in 2014, 10,213 in 2011 and 10,051 in 2008. Approximately 72% of the geoscience faculty is tenured and 15% are untenured but in tenure-track positions (Figure 3.2).

The age distribution of geoscience faculty has shifted to an average age of 58 years old in 2015. The ages of the majority of faculty ranked as professor remained steady at 61-65 years of age since 2014 (Figure 3.3). There was also an increase in the number of assistant professors from 1,125 faculty members in 2013 to 1,142 faculty members in 2015 indicating a small increase in hiring for tenure-track positions at four-year universities over the past two years. The number of emeritus faculty has decreased to 1,000 people, compared to 1,160 people in 2014.

The overall percentage of women in geoscience faculty positions at four-year universities continues to slowly rise from 17% in 2013 to 19% in 2015. This percentage in women geoscience faculty has been consistently increasing since 2008. This increase is also seen among all the different ranks given to faculty. The assistant professor and instructor/lecturer positions continue to have the highest percentage of women faculty members (Figure 3.4).

According to AGI's GeoRef database, over the past several decades, there has been a 128% increase in the number of publications produced in geology, from 161,506 in the 1970's to 368,885 in the 2000-2013 (Figure 3.5). There have been rapid increases in the number of publications in the environmental geology, quaternary geology, geophysics, and economic geology subject areas in particular. In the past few decades, economic geology had the highest number of publications, but recently environmental geology rapidly grew in the past decade to the highest number of publications. Publication rates within journals sponsored by geoscience societies show increases in the number of articles published in the atmospheric sciences, space physics, and climatology in particular (Figure 3.6). However, in 2015, there was a decrease in the annual number of publications in atmospheric sciences published in JGR Atmospheres.

Since 1999, there has been a slight increase in the number of faculty specializing in other geoscience fields, such as the atmospheric sciences, earth science education, and geographic information systems fields (Figure 3.7). Some of these fields within the "other" category may need to be moved out of this category as some of these areas become more popular specializations.

Figure 3.2: Percentage of Geoscience Faculty by Rank at Four-Year Institutions, 2015

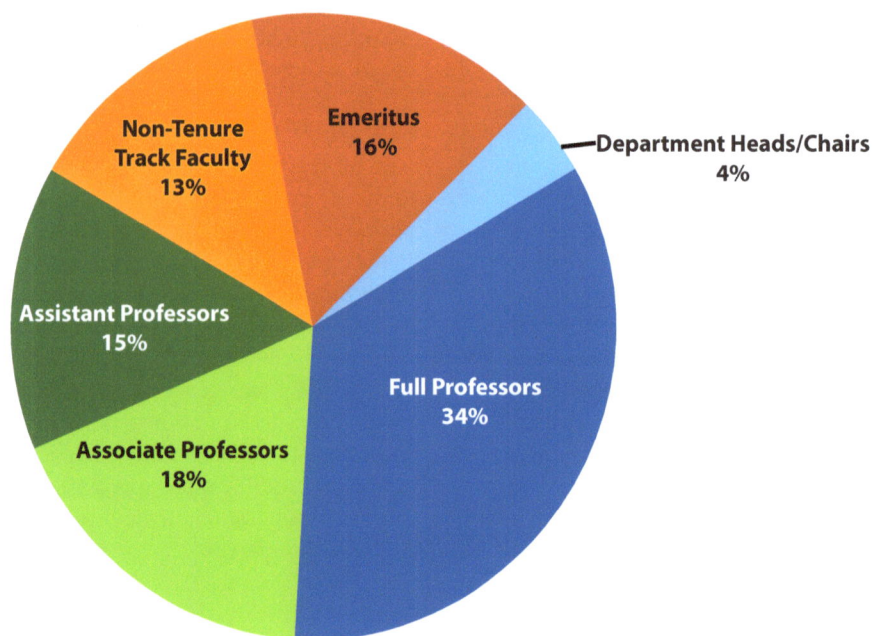

AGI Geoscience Workforce Program; Data derived from AGI's Directory of Geoscience Departments database

Table 3.4: Top Ten Degree Granting Institutions of U.S. Geoscience Tenure-Track or Tenured Faculty, 2015

School where Faculty Earned Highest Degree	Total Number of Tenure or Tenure-Track Faculty Graduates
Massachusetts Institute of Technology	293
University of Washington	230
University of California-Berkeley	225
University of Wisconsin-Madison	214
Stanford University	210
Harvard University	176
California Institute of Technology	172
Columbia University	171
Pennsylvania State University	164
University of Arizona	157

AGI Geoscience Workforce Program; Data derived from AGI's Directory of Geoscience Departments database

Figure 3.3: Number of Geoscience Faculty by Age Group and Rank, 2015

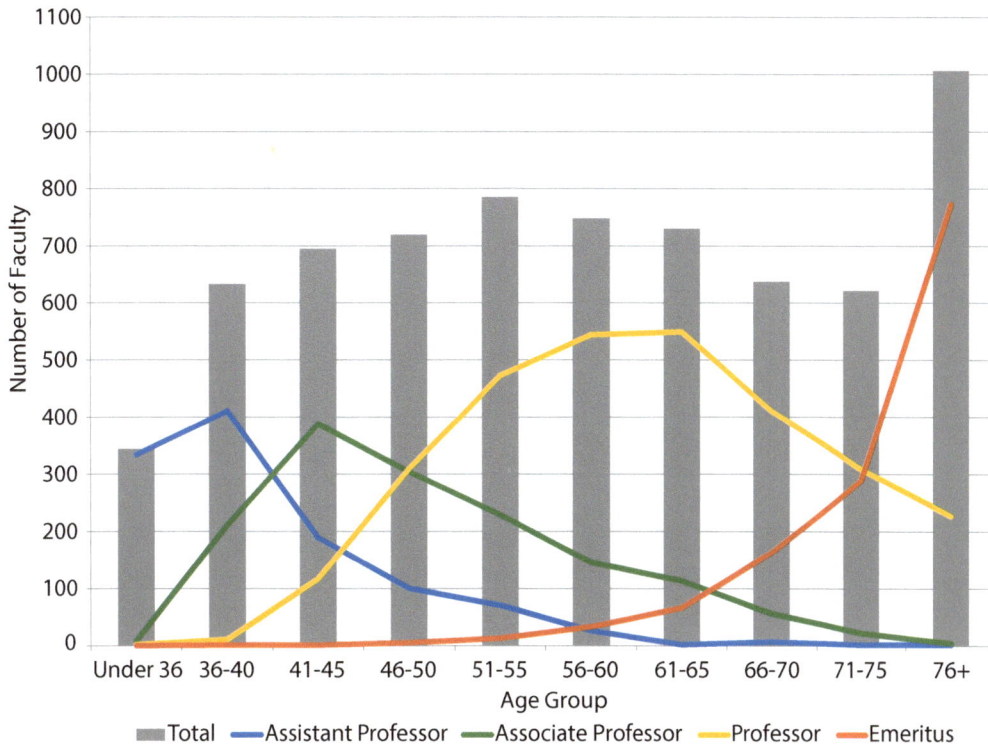

AGI Geoscience Workforce Program; Data derived from AGI's Directory of Geoscience Departments database

Figure 3.4: Percentage of Female Geoscience Faculty by Rank

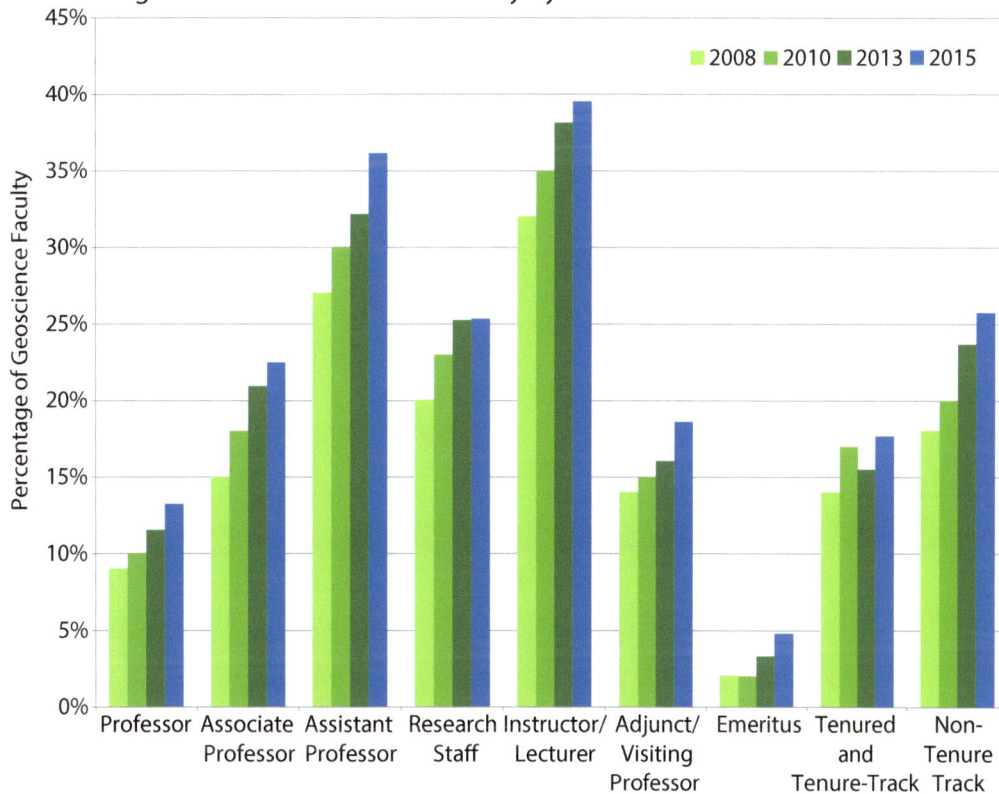

AGI Geoscience Workforce Program; Data derived from AGI's Directory of Geoscience Departments database

Figure 3.5: Trends in Geoscience Publications

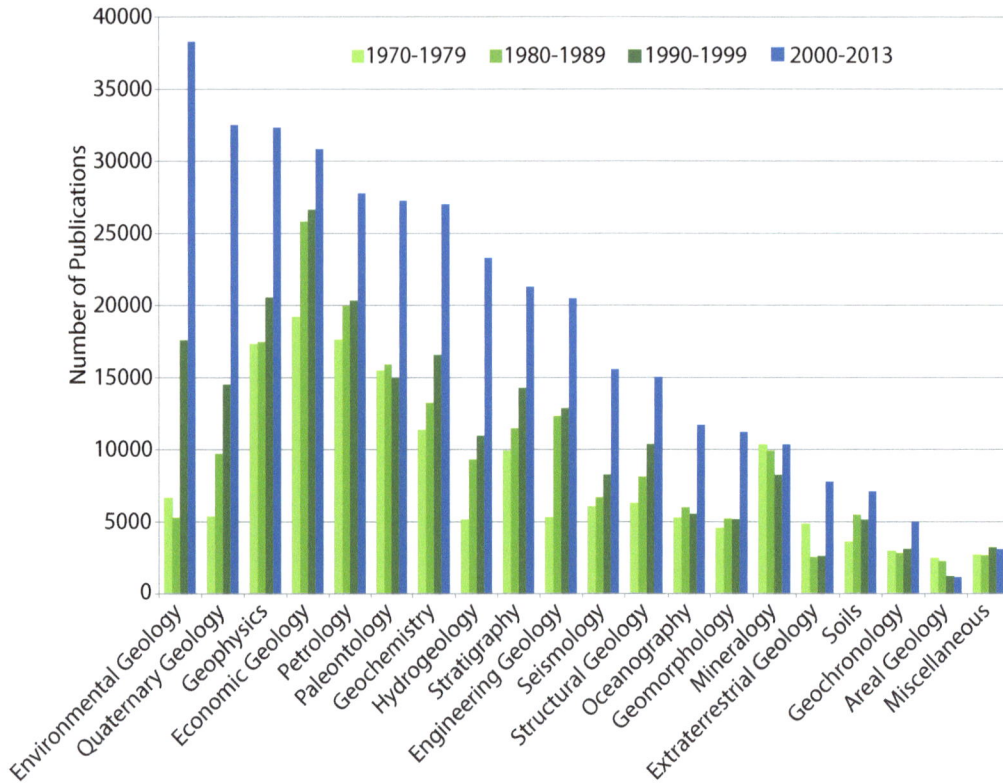

AGI Geoscience Workforce Program; Data derived from AGI's Georef database

Table 3.5: Top Five Geoscience Publication Topics, 1970-2013

1970-1979	1980-1989	1990-1999	2000-2013
Economic Geology	Economic Geology	Economic Geology	Environmental Geology
Petrology	Petrology	Geophysics	Quaternary Geology
Geophysics	Geophysics	Petrology	Geophysics
Paleontology	Paleontology	Environmental Geology	Economic Geology
Geochemistry	Geochemistry	Geochemistry	Petrology

AGI Geoscience Workforce Program; Data derived from AGI's GeoRef database

Figure 3.6: Publication Trends in Selected Geoscience Journals

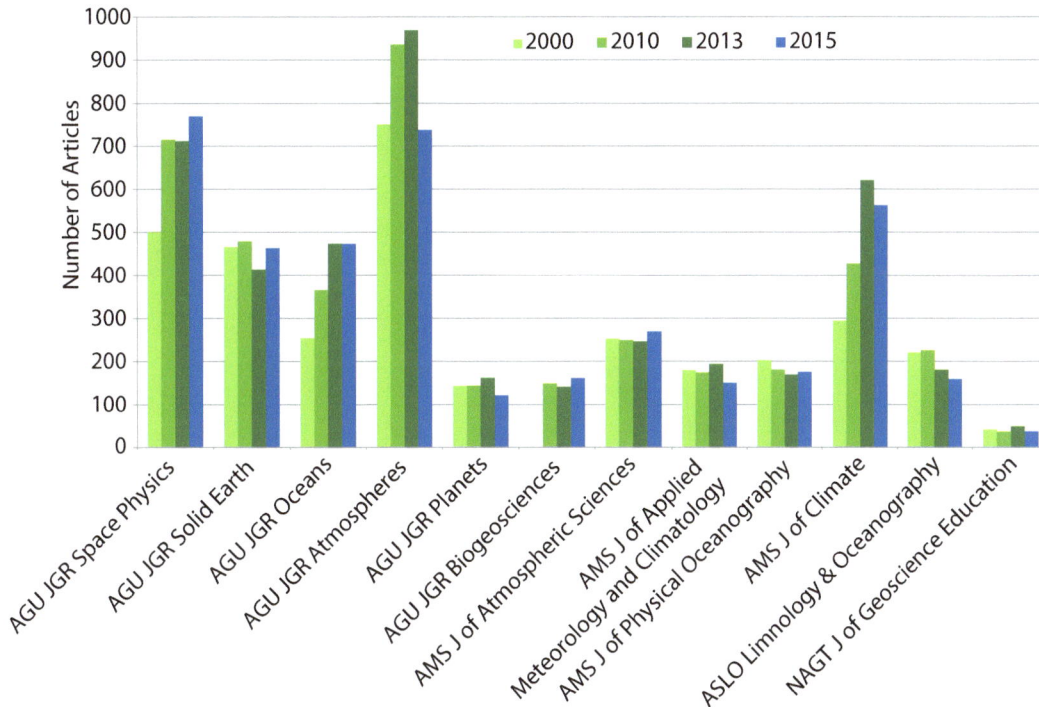

Acromyns: American Geophysical Union (AGU) Journal of Geophysical Research (JGR); American Meteorological Union (AMS); Association for the Sciences of Limnology and Oceanography (ASLO); National Association of Geoscience Teachers (NAGT)

AGI Geoscience Workforce Program; Data derived from the AGU, AMS, ASLO, and NAGT journal plublication websites

Figure 3.7: Trends in Geoscience Faculty Specialities (1999-2015)

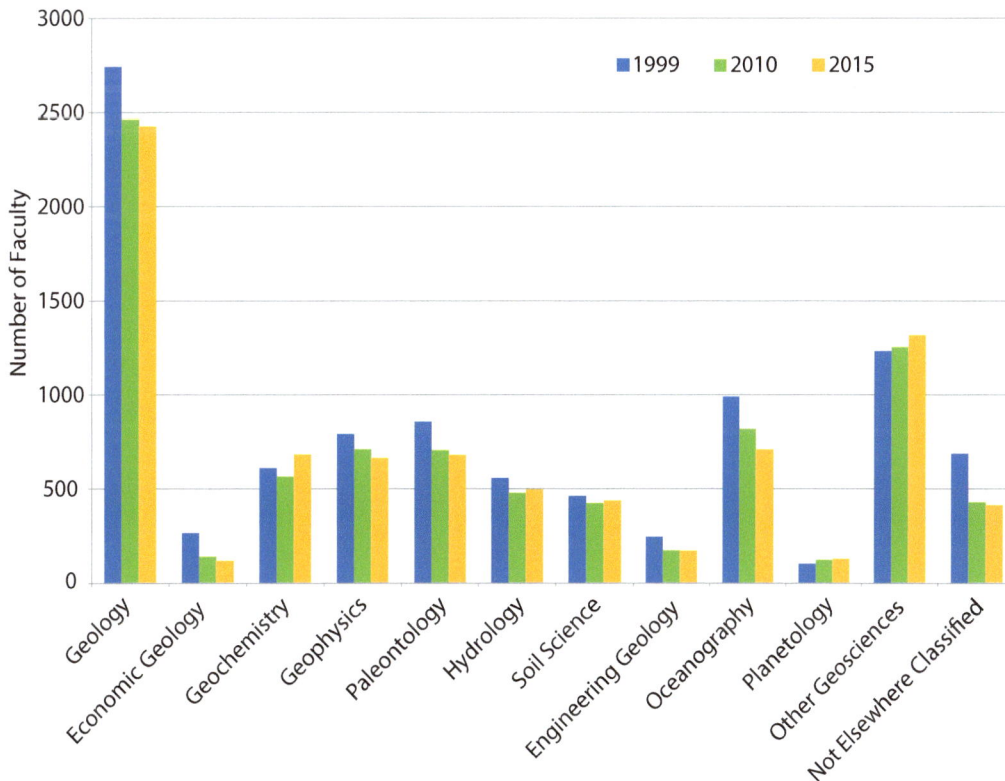

AGI Geoscience Workforce Program; Data derived from AGI's Directory of Geoscience Departments database

Table 3.6: Research and Teaching Specialties of Geoscience Faculty, 2015

Faculty Specialties	Faculty (2015)	Faculty Specialties	Faculty (2015)	Faculty Specialties	Faculty (2015)
GEOLOGY		Marine Geochemistry	58	Soil Biology/Biochemistry	54
General Geology	216	Organic Geochemistry	54	Paleopedology/Archeology	6
Archaeological Geology	42	Stable Isotopes	87	Other Soil Science	111
Environmental Geology	114	Trace Element Distribution	16	**ENGINEERING GEOLOGY**	
Geomorphology	245	**GEOPHYSICS**		General Engineering Geology	61
Glacial Geology	77	General Geophysics	190	Earthquake Engineering	4
Marine Geology	83	Experimental Geophysics	27	Mining Tech/Extractive Metallurgy	14
Mineralogy and Crystallography	135	Exploration Geophysics	77	Mining Engineering	39
Paleolimnology	18	Geodesy	20	Petroleum Engineering	23
Petroleum Geology	51	Geomagnetism and Paleomagnetism	51	Rock Mechanics	28
General Petrology	81	Gravity	6	**OCEANOGRAPHY**	
Igneous Petrology	206	Heat Flow	10	General Oceanography	50
Metamorphic Petrology	90	Seismology	209	Biological Oceanography	230
Sedimentary Petrology	67	Marine Geophysics	73	Chemical Oceanography	120
Sedimentology	309	**PALEONTOLOGY**		Geological Oceanography	76
Physical Stratigraphy	89	General Paleontology	88	Physical Oceanography	187
Structural Geology	330	Paleostratigraphy	41	Shore and Nearshore Processes	45
Tectonics	124	Micropaleontology	59	**PLANETOLOGY**	
Volcanology	80	Paleobotany	25	Cosmochemistry	20
Mathematical Geology	21	Palynology	22	Extraterrestrial Geology	56
Mineral Physics	27	Quantitative Paleontology	2	Extraterrestrial Geophysics	38
History of Geology	6	Vertebrate Paleontology	103	Meteorites and Tektites	14
Geomedicine	3	Invertebrate Paleontology	102	**OTHER**	
Forensic Geology	0	Paleobiology	71	General Earth Sciences	68
ECONOMIC GEOLOGY		Paleoecology and Paleoclimatology	111	Atmospheric Sciences	506
General Economic Geology	56	Geobiology	52	Earth Science Education	75
Coal	15	**HYDROLOGY**		Physical Geography	127
Metals	35	General Hydrology	109	Ocean Engineering/Mining	9
Non-Metals	2	Ground Water/Hydrogeology	261	Remote Sensing	130
Oil and Gas	9	Quantitative Hydrology	40	Soil Science	31
GEOCHEMISTRY		Surface Water	49	Meteorology	147
General Geochemistry	160	Geohydrology	35	Material Science	18
Analytical Geochemistry	31	**SOIL SCIENCE**		Land Use/Urban Geology	50
Experimental Petrology/Phase Equilibria	42	Soil Physics/Hydrology	65	Geographic Information Systems	138
Exploration Geochemistry	6	Soil Chemistry/Mineralogy	116	Glaciology	16
Geochronology and Radioisotopes	86	Pedology/Classification/Morphology	54	Not Elsewhere Classified	411
Low-Temperature Geochemistry	141	Forest Soils/Rangelands/Wetlands	29		

AGI Geoscience Workforce Program; Data derived from AGI's Directory of Geoscience Departments database

Table 3.7: Top Geoscience Specialities with the Most Change in Faculty Since 2013

Positive Change in Faculty	Negative Change in Faculty
GEOLOGY	
General Geology	Metamorphic Geology
Volcanology	Structural Geology
Petroleum Geology	Sedimentary Petrology
ECONOMIC GEOLOGY	
Oil and Gas	Metals
--	General Economic Geology
GEOCHEMISTRY	
General Geochemistry	Marine Geochemistry
Stable Isotopes	Experimental Petrology
GEOPHYSICS	
Seismology	Marine Geophysics
Geodesy	Exploration Geophysics
PALEONTOLOGY	
Paleobiology	Invertebrate Paleontology
--	Paleoecology and Paleoclimatology
HYDROLOGY	
General Hydrology	--
Quantitative Hydrology	--
SOIL SCIENCE	
Other Soil Science	Soil Chemistry/Mineralogy
Soil Physics/Hydrology	Soil Biology/Biochemistry
ENGINEERING GEOLOGY	
Mining Engineering	Earthquake Engineering
Petroleum Engineering	--
OCEANOGRAPHY	
Chemical Oceanography	Biological Oceanography
General Oceanography	Physical Oceanography
PLANETOLOGY	
Extraterrestrial Geology	Cosmochemistry
--	Extraterrestrial Geophysics
OTHER GEOSCIENCES	
Atmospheric Sciences	Meteorology
Geographic Information Sciences	Physical Geography

AGI Geoscience Workforce Program; Data derived from AGI's Directory of Geoscience Departments database

Geoscience University Students

While enrollments at four-year universities increased by 39% between 2001 and 2011 to 13,494,131 students, enrollments changed very little from 2011 to 2013. This was a similar trend seen in two-year college enrollments, indicating some trend or issue affecting enrollments for postsecondary degrees. While the predicted enrollments at four-year universities trend upwards, this trend may not be accurate because it does not fully take into account the relatively stagnant growth in enrollments over the past few years. Enrollments of female students at four-year universities have hovered at 57% of the total enrollments since 2005. The percentage of women enrolled at four-year universities is about the same as the percentage of women enrolled at two-year colleges. Also as in two-year colleges, the overall percentages of underrepresented minorities enrolled at four-year universities continue to increase. However, the percentage of African Americans enrolled at four-year universities decreased slightly from 14% in 2010 to 13% in 2013.

For the geosciences, undergraduate enrollments were increasing steadily from 2007 to 2015 reaching 31,219 students. While there was a small drop in enrollments in 2013, this general increasing trend in enrollments in 2011 and 2012 go against the near stagnant change in overall enrollments at four-year universities. Whatever is affecting enrollments in postsecondary degrees may not be affecting enrollments into geoscience departments. Graduate enrollments continued to fluctuate around 10,000 students each year. Geoscience degrees awarded for bachelor's students were trending upward from 2006-2013, but awarded bachelor's degrees decreased from 4,146 to 3,629 in 2015. Degrees awarded to master's students were at 1,511 in 2012, but from 2013-2015, master's graduation rates hovered just under 1500. Doctorates awarded hovered between 600-700 students a year, but in 2015, awarded doctorates fell below 600 to 597.

Enrollments in the geosciences among women have continued to decrease, and awarded geoscience degrees to women have fluctuated between 40-45%. The recent increases in enrollments have been largely driven by male students.

Figures 3.14-3.18 show the percentages of science and geoscience degrees awarded to underrepresented minorities from three different sources—the National Science Foundation, the Department of Education's IPEDS database, and AGI's Geoscience Student Exit Survey. Accurate data on the enrollments and completions of underrepresented minorities in geoscience degree programs can be difficult to acquire. Figures 3.16-3.18 show the data provided by the Department of Education. Three sources are used because each has slightly different methods to collect demographic information. The Department of Education IPEDS database contains information received from the main administrative offices at universities, so their information comes from the forms filled out by the students upon entry into the university and any supplemental data provided by the departments. This can affect the accuracy of the information of race and ethnicity of geoscience graduates, especially since the percentages tend to be fairly low for this group of students. The data collected by the National Science Foundation tends to come from the institutions, much like the Department of Education, and in the case of doctorates, the students directly. AGI's Geoscience Student Exit Survey collects data directly from the recent graduates at the time of their graduation. Also, all three data sets represent data from different years ranging from 2013-2015.

Not surprisingly, the majority of geoscience students are U.S. citizens. Geoscience departments begin to see a larger population of non-permanent residents in graduate programs, particularly doctoral programs, where over a quarter of the graduates are from a different country (Figure 3.19).

Concern has been raised recently about the overall trends of the socioeconomic status of geoscience students, and one way to infer this is through the highest education level of the students' parents or guardians. In 2015, this question was added to AGI's Geoscience Student Exit Survey, and the data is presented in Figure 3.21. In general, most geoscience graduates at all degree levels have at least one parent with a postsecondary degree, and the percentage of graduates with a parent holding a graduate or professional degree increases with the graduate's degree level. Also, this data indicated 12% of bachelor's graduates, 10% of master's graduates, and 13% of doctoral graduates were first generation college students.

While geoscience students at four-year universities are working toward their degrees, nearly 80% of all geoscience graduates complete Calculus I, but the percentages of bachelor's and master's graduates with experience in higher-level quantitative courses drops dramatically after Calculus II. Students are receiving instruction in chemistry and physics during their geoscience degrees as well, with 77-90% of graduates at the differing degree-levels taking at least one course in chemistry and approximately

90% of all graduates taking calculus or algebra-based physics (Figure 3.22).

In 2015, the majority of graduates at the bachelor's and master's degree levels chose to major in the geosciences at some point during their undergraduate education. This trend has also been seen in previous years through AGI's Geoscience Student Exit Survey, which highlights the importance of undergraduate geoscience courses as recruitment tools for future majors. Most doctoral graduates in 2015 indicated choosing to major in the geosciences either after receiving an undergraduate degree or during their undergraduate education (Figure 3.23). Figure 3.24 displays the chosen fields for their geoscience degrees in 2015. This figure highlights the varied fields and interests that fall under the geoscience umbrella.

Figure 3.8: Fall Enrollments at Four-Year Institutions

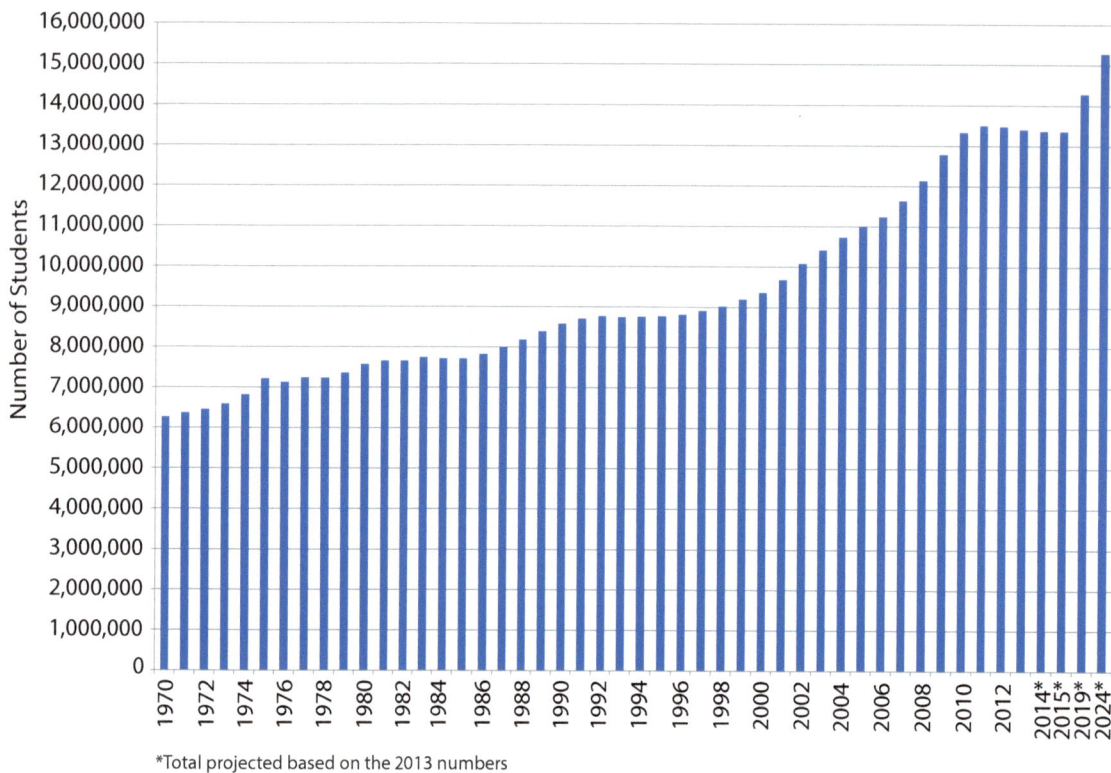

*Total projected based on the 2013 numbers

AGI Geoscience Workforce Program; Data derived from NCES Digest of Education Statistics, 2014

Figure 3.9: Participation of Women in Four-Year Institutions

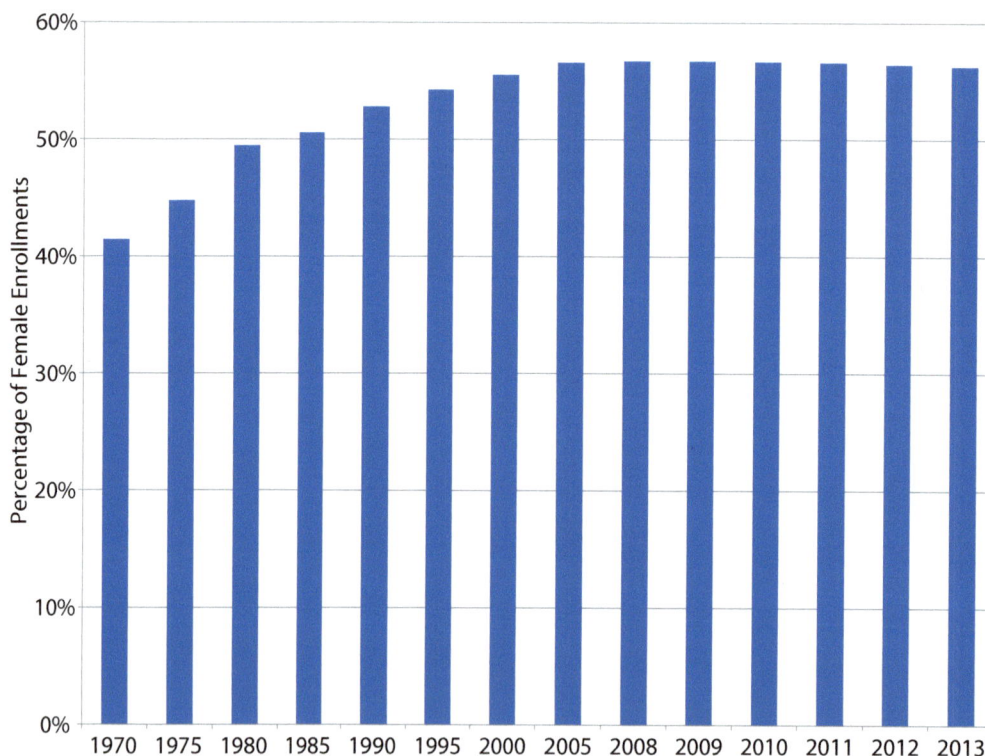

AGI Geoscience Workforce Program; Data derived from NCES Digest of Education Statistics, 2014

Figure 3.10: Underrepresented Minority Enrollments at Four-Year Institutions

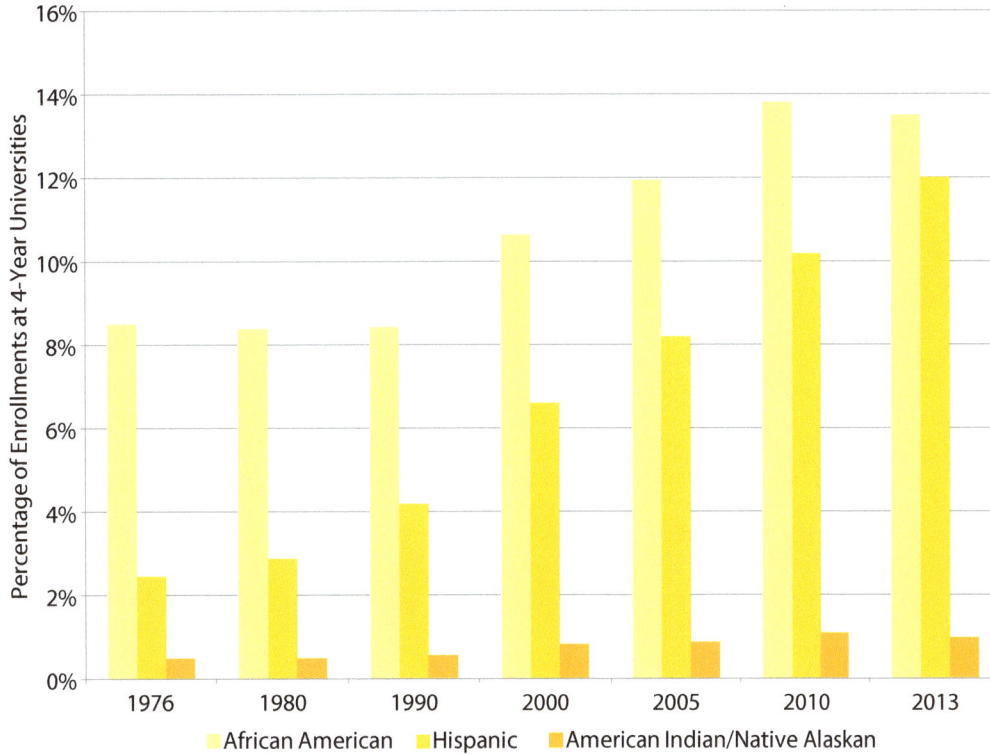

AGI Geoscience Workforce Program; Data derived from NCES Digest of Education Statistics, 2014

Figure 3.11: Geoscience Enrollments at U.S. Four-Year Institutions, 1955-2015

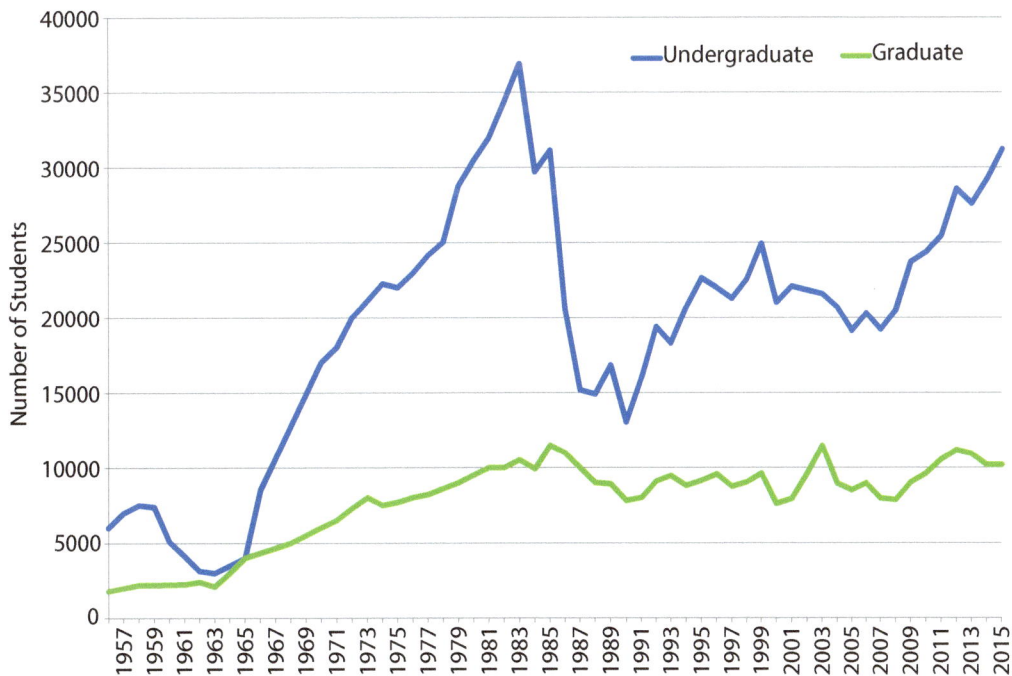

AGI Geoscience Workforce Program; Data derived from AGI's Directory of Geoscience Departments database

Figure 3.12: Geoscience Degrees Awarded at U.S. Four-Year Institutions, 1973-2015

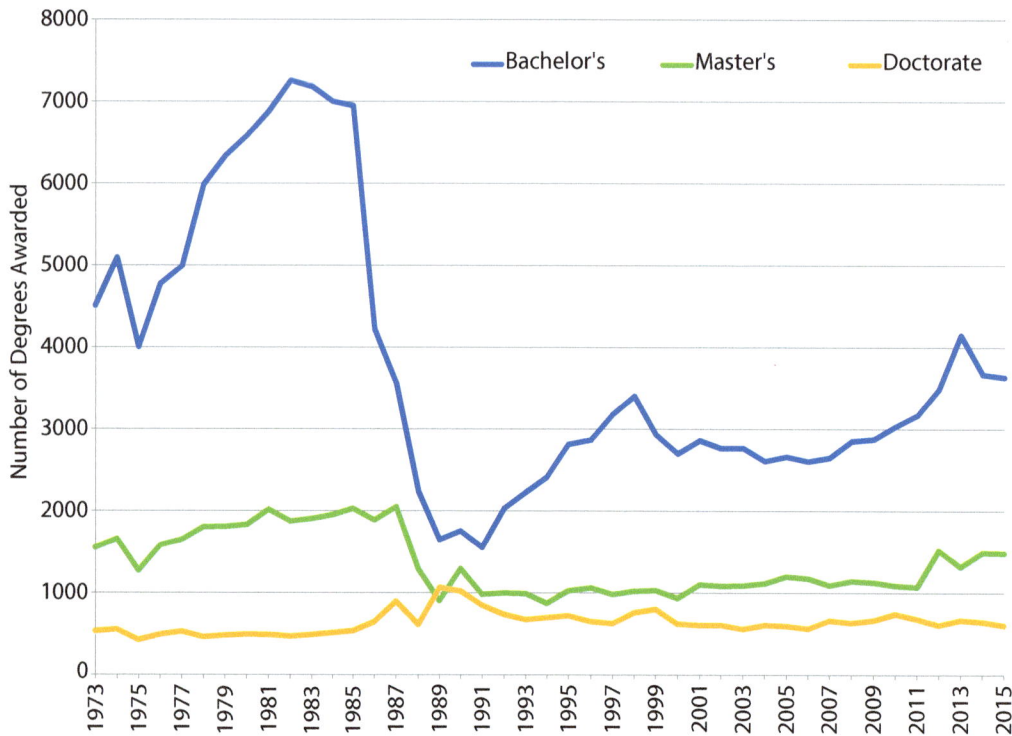

AGI Geoscience Workforce Program; Data derived from AGI's Directory of Geoscience Departments database

Figure 3.13: Participation of Women in Geoscience Programs

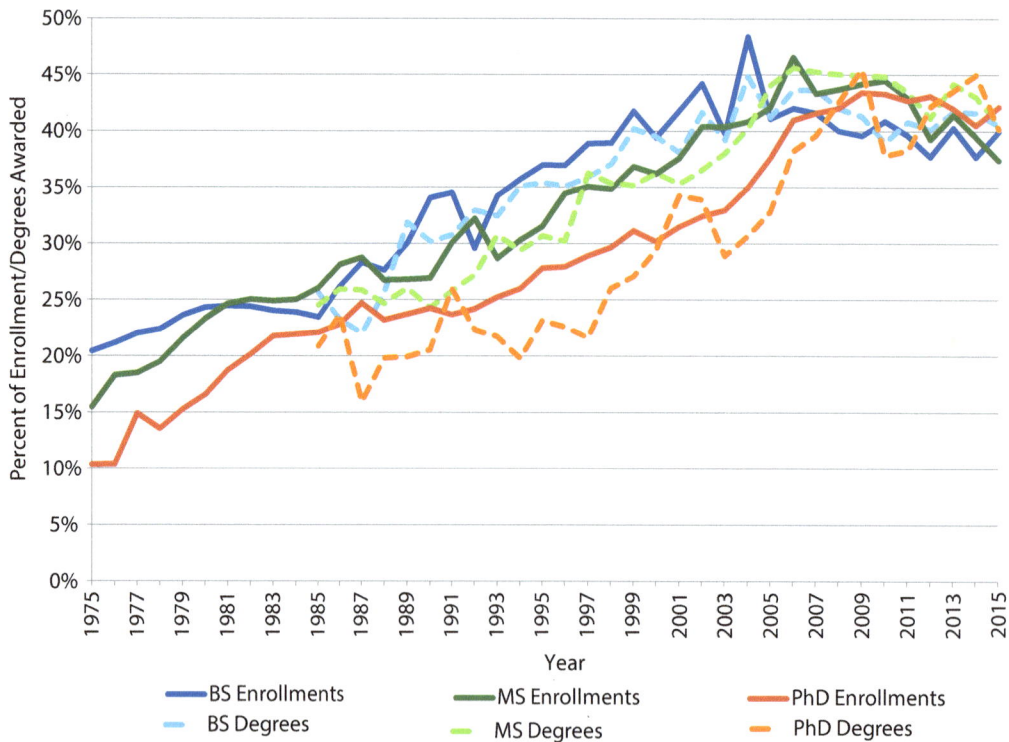

AGI Geoscience Workforce Program; Data derived from AGI's Directory of Geoscience Departments database

Figure 3.14: Percentage of Science and Engineering Degrees Awarded to Underrepresented Minorities, 2013

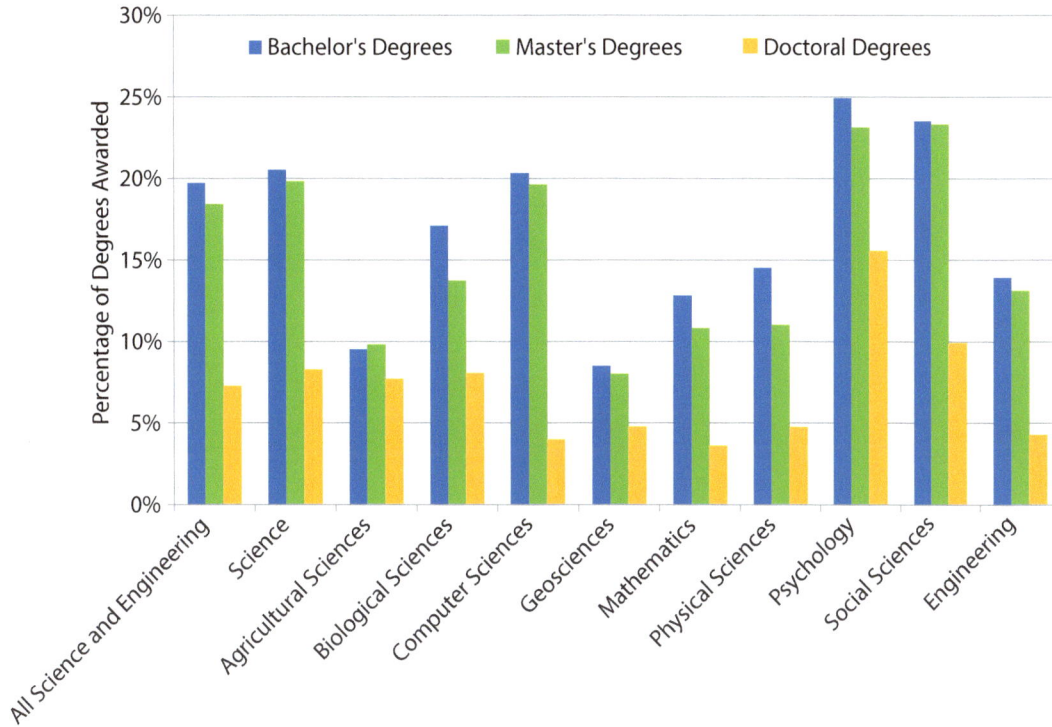

AGI Geoscience Workforce Program; Data derived from NSF's SESTAT Restricted-Use Data files, 2013

Figure 3.15: Percentage of Geoscience Bachelor's Degrees Awarded to Underrepresented Minorities

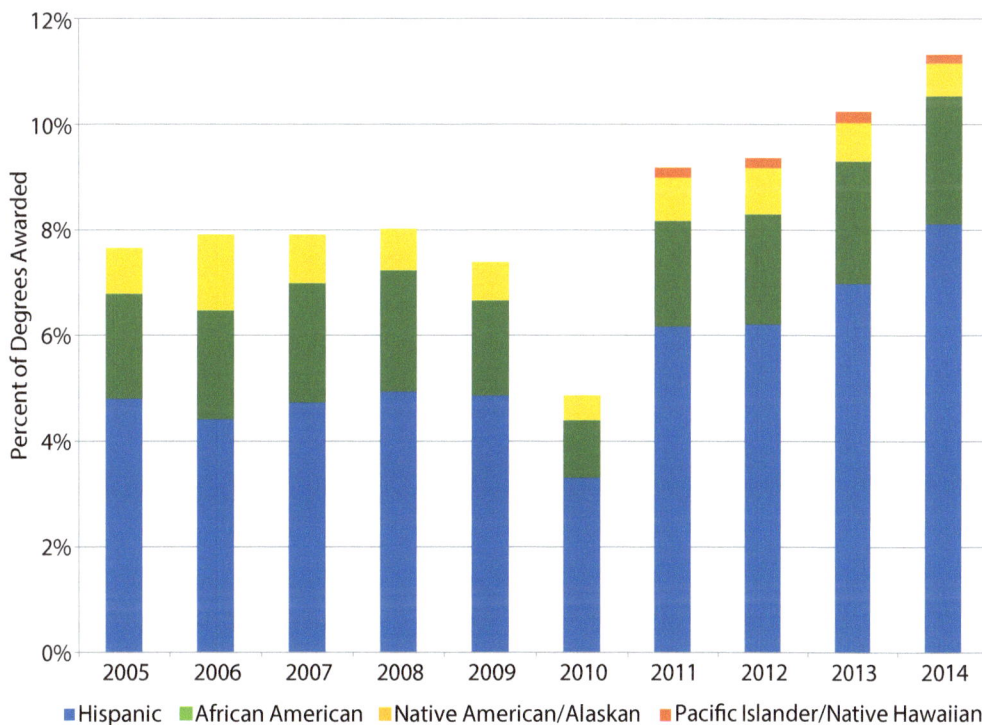

AGI Geoscience Workforce Program; Data derived from IPEDS

Figure 3.16: Percentage of Geoscience Master's Degrees Awarded to Underrepresented Minorities

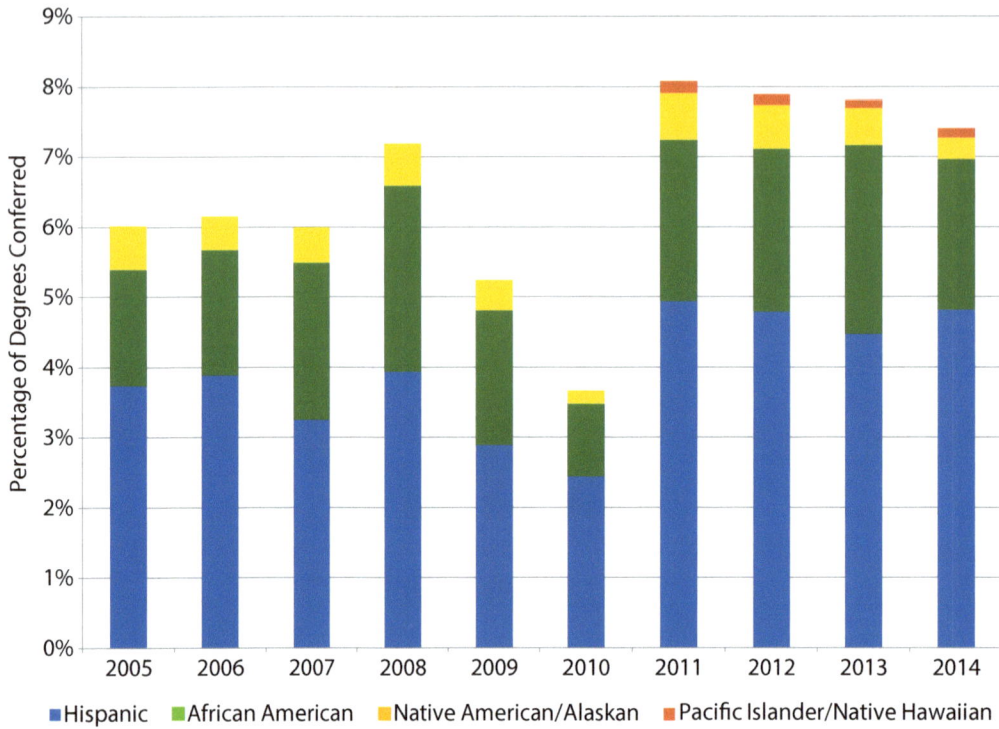

AGI Geoscience Workforce Program; Data derived from IPEDS

Figure 3.17: Percentage of Geoscience Doctoral Degrees Awarded to Underrepresented Minorities

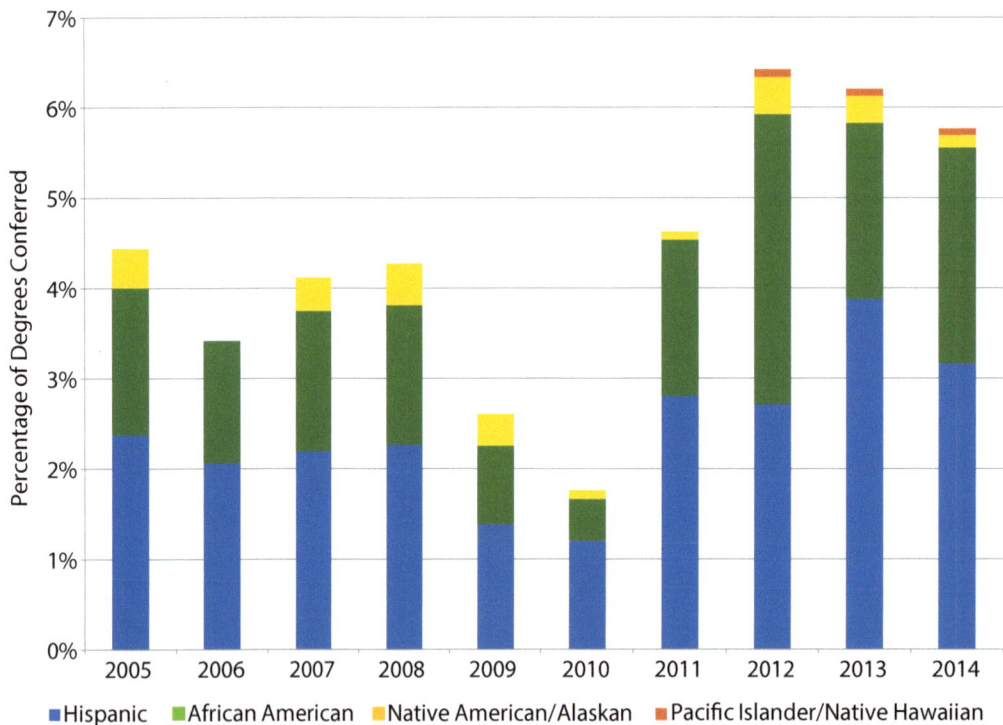

AGI Geoscience Workforce Program; Data derived from IPEDS

Figure 3.18: Race and Ethnicity of Geoscience Graduates, 2015

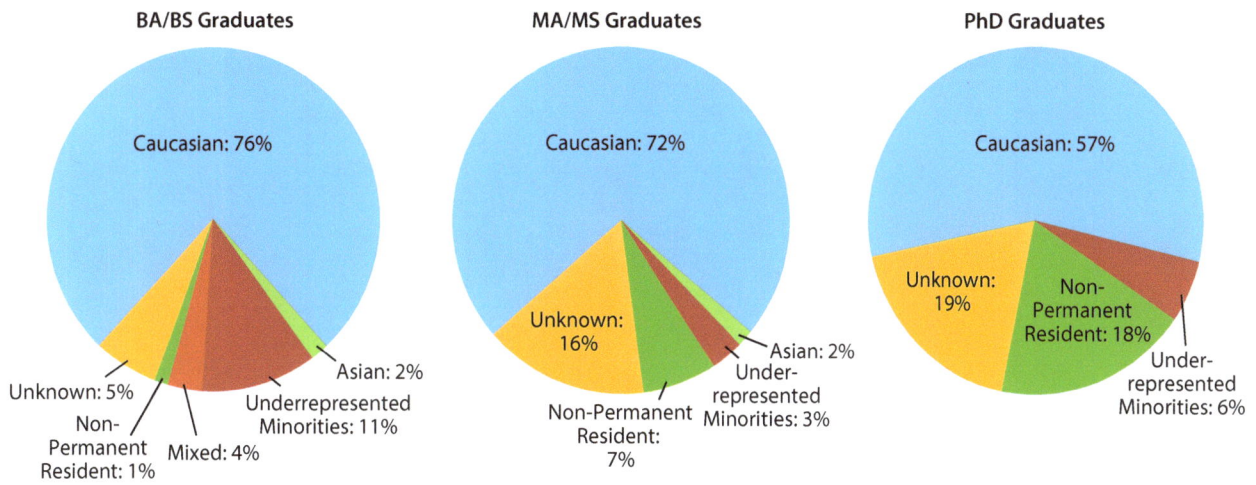

BA/BS Graduates

Caucasian: 76%

Unknown: 5%

Non-Permanent Resident: 1%

Mixed: 4%

Underrepresented Minorities: 11%

Asian: 2%

MA/MS Graduates

Caucasian: 72%

Unknown: 16%

Non-Permanent Resident: 7%

Under-represented Minorities: 3%

Asian: 2%

PhD Graduates

Caucasian: 57%

Unknown: 19%

Non-Permanent Resident: 18%

Under-represented Minorities: 6%

AGI Geoscience Workforce Program; Data derived from AGI's Geoscience Student Exit Survey 2015

Figure 3.19: Citizenship of Geoscience Graduates, 2015

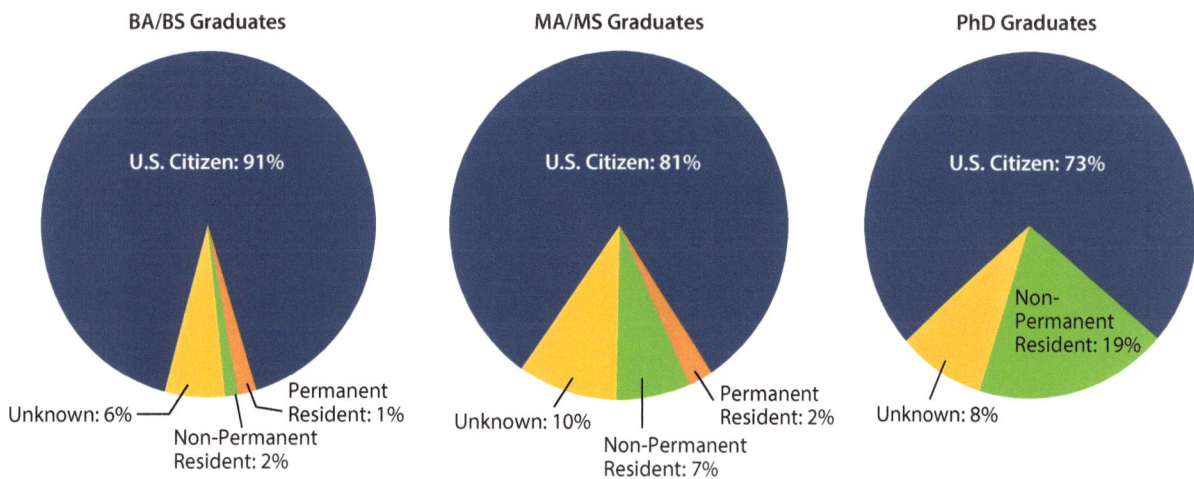

BA/BS Graduates

U.S. Citizen: 91%

Unknown: 6%

Non-Permanent Resident: 2%

Permanent Resident: 1%

MA/MS Graduates

U.S. Citizen: 81%

Unknown: 10%

Non-Permanent Resident: 7%

Permanent Resident: 2%

PhD Graduates

U.S. Citizen: 73%

Non-Permanent Resident: 19%

Unknown: 8%

AGI Geoscience Workforce Program; Data derived from AGI's Geoscience Student Exit Survey 2015

Figure 3.20: Highest Level of Education of Parents of Geoscience Graduates, 2015

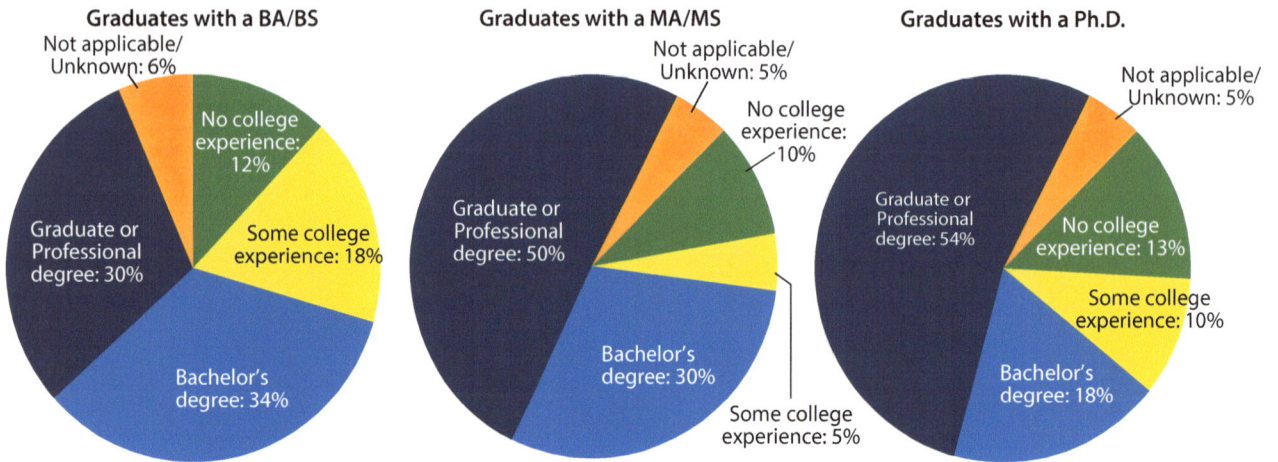

Graduates with a BA/BS

Not applicable/
Unknown: 6%

No college
experience:
12%

Some college
experience: 18%

Graduate or
Professional
degree: 30%

Bachelor's
degree: 34%

Graduates with a MA/MS

Not applicable/
Unknown: 5%

No college
experience:
10%

Graduate or
Professional
degree: 50%

Bachelor's
degree: 30%

Some college
experience: 5%

Graduates with a Ph.D.

Not applicable/
Unknown: 5%

No college
experience: 13%

Some college
experience: 10%

Graduate or
Professional
degree: 54%

Bachelor's
degree: 18%

AGI Geoscience Workforce Program; Data derived from AGI's Geoscience Student Exit Survey 2015

Figure 3.21: Quantitative Skills and Knowledge Gained by Geoscience Graduates, 2015

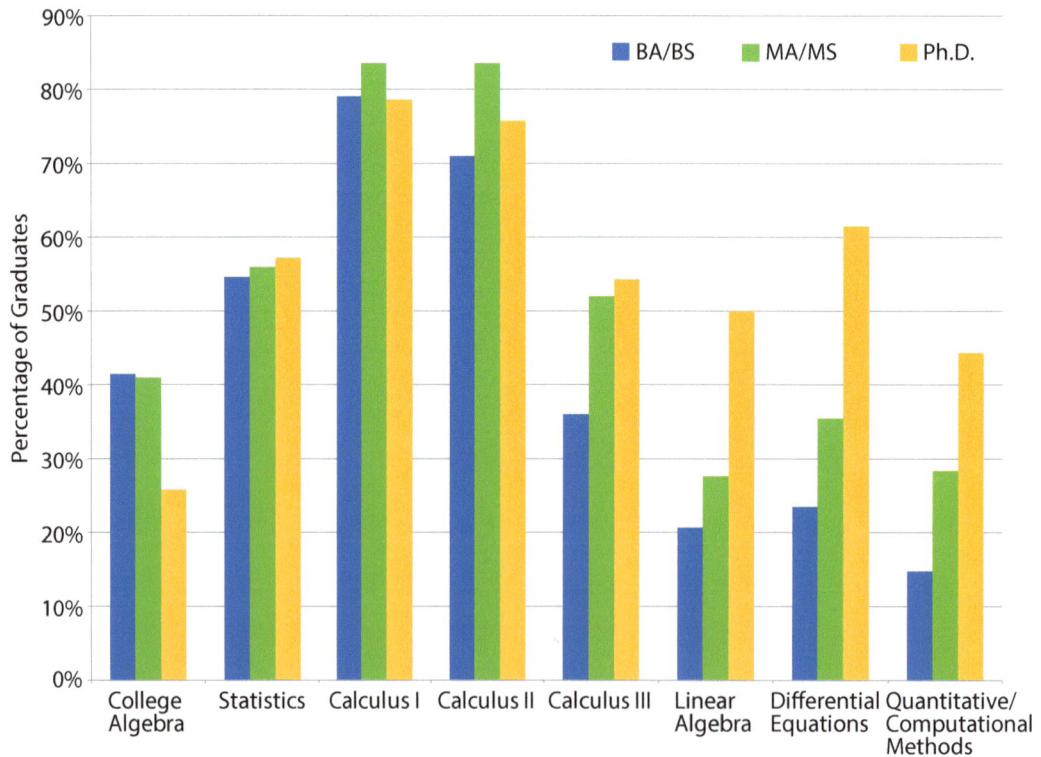

AGI Geoscience Workforce Program; Data derived from AGI's Geoscience Student Exit Survey 2015

Figure 3.22: Supplemental Science Courses Taken by Geoscience Graduates, 2015

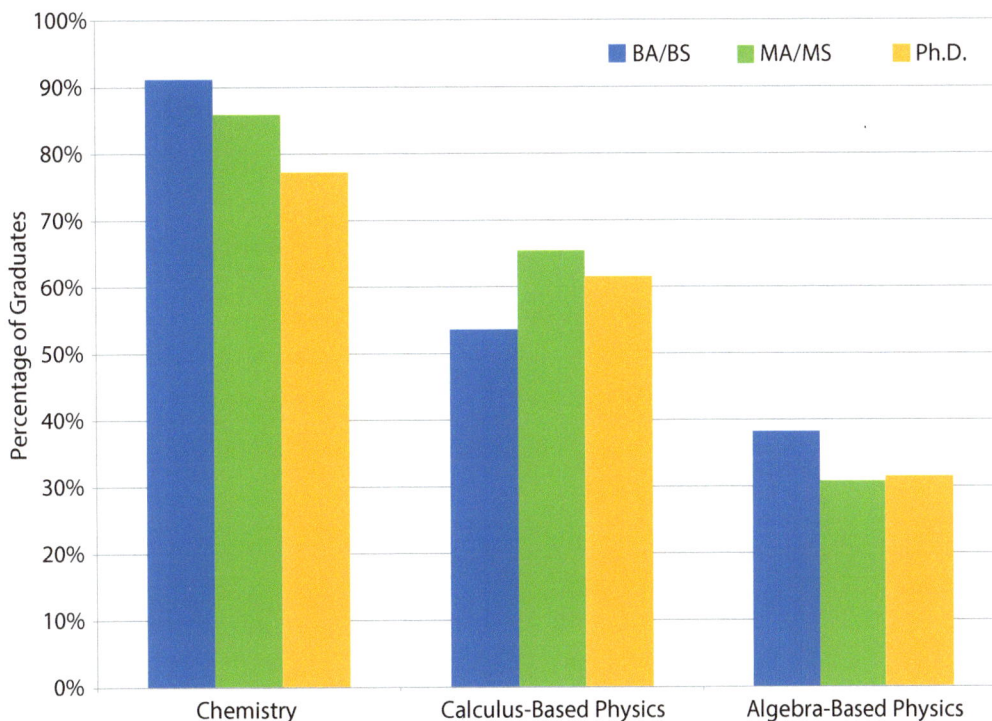

AGI Geoscience Workforce Program; Data derived from AGI's Geoscience Student Exit Survey 2015

Figure 3.23: The Point in Time When Geoscience Graduates Decided to Major in the Geosciences, 2015

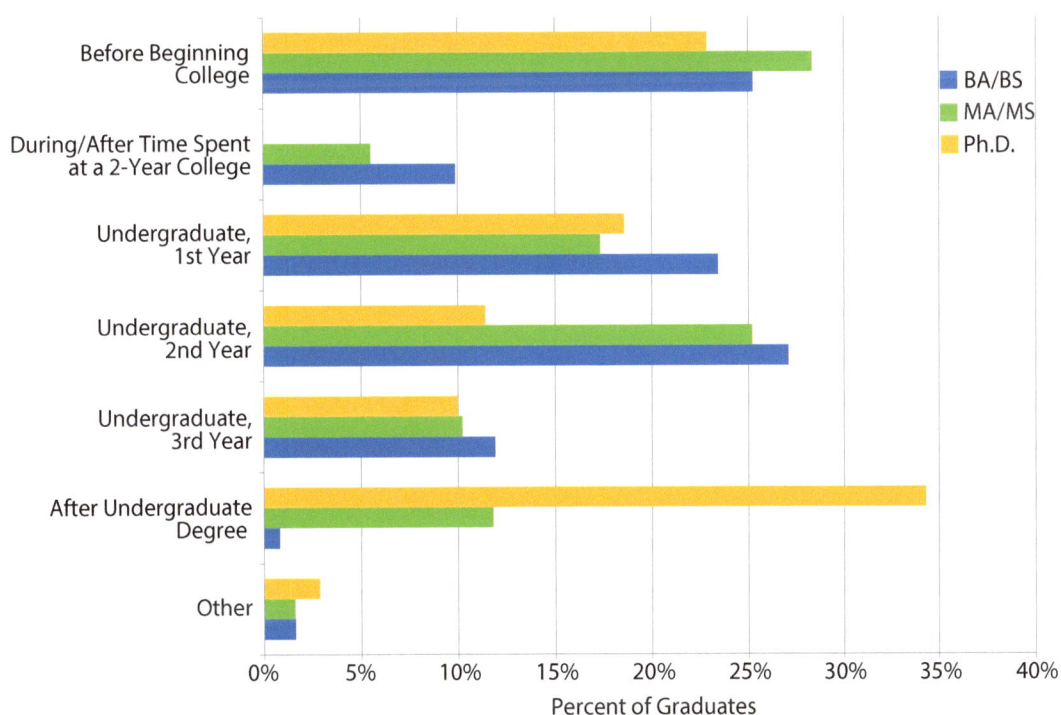

AGI Geoscience Workforce Program; Data derived from AGI's Geoscience Student Exit Survey 2015

Figure 3.24: The Chosen Degree Fields of Geoscience Graduates, 2015

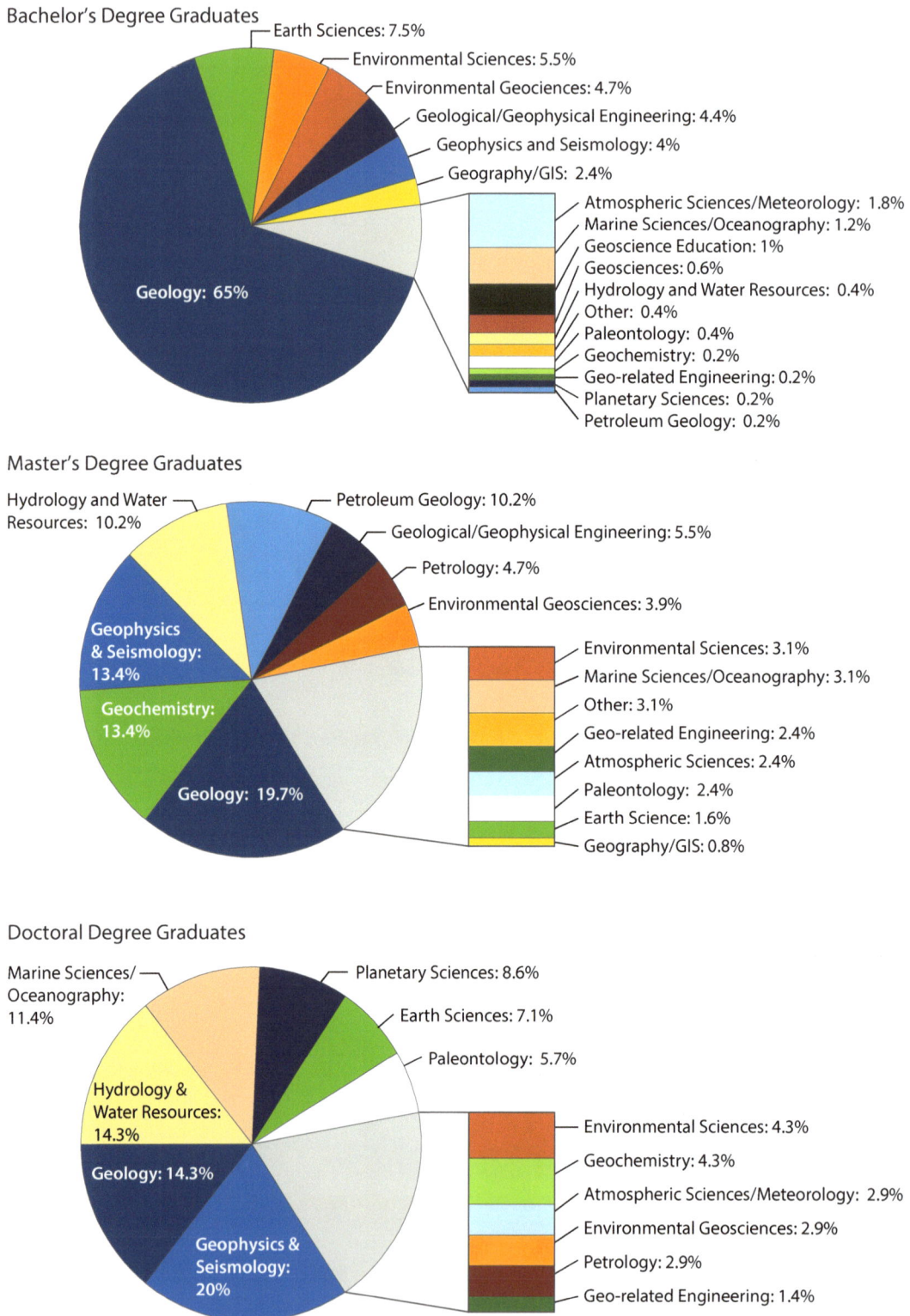

Bachelor's Degree Graduates

Earth Sciences: 7.5%
Environmental Sciences: 5.5%
Environmental Geociences: 4.7%
Geological/Geophysical Engineering: 4.4%
Geophysics and Seismology: 4%
Geography/GIS: 2.4%
Geology: 65%

Atmospheric Sciences/Meteorology: 1.8%
Marine Sciences/Oceanography: 1.2%
Geoscience Education: 1%
Geosciences: 0.6%
Hydrology and Water Resources: 0.4%
Other: 0.4%
Paleontology: 0.4%
Geochemistry: 0.2%
Geo-related Engineering: 0.2%
Planetary Sciences: 0.2%
Petroleum Geology: 0.2%

Master's Degree Graduates

Hydrology and Water Resources: 10.2%
Petroleum Geology: 10.2%
Geological/Geophysical Engineering: 5.5%
Petrology: 4.7%
Environmental Geosciences: 3.9%
Geophysics & Seismology: 13.4%
Geochemistry: 13.4%
Geology: 19.7%

Environmental Sciences: 3.1%
Marine Sciences/Oceanography: 3.1%
Other: 3.1%
Geo-related Engineering: 2.4%
Atmospheric Sciences: 2.4%
Paleontology: 2.4%
Earth Science: 1.6%
Geography/GIS: 0.8%

Doctoral Degree Graduates

Marine Sciences/Oceanography: 11.4%
Planetary Sciences: 8.6%
Earth Sciences: 7.1%
Paleontology: 5.7%
Hydrology & Water Resources: 14.3%
Geology: 14.3%
Geophysics & Seismology: 20%

Environmental Sciences: 4.3%
Geochemistry: 4.3%
Atmospheric Sciences/Meteorology: 2.9%
Environmental Geosciences: 2.9%
Petrology: 2.9%
Geo-related Engineering: 1.4%

AGI Geoscience Workforce Program; Data derived from AGI's Geoscience Student Exit Survey 2015; Figure created by Kathleen Cantner

Co-Curricular Activities

Field camp attendance hit a high of 3,237 student participants in 2014, but the attendance dropped in 2015 to 2,867 students (Figure 3.25). If this trend continues, then either the available field camps have reached their annual capacity or the cost of attending field camp may be prohibiting participation among students. Most states in the United States either have at least one field camp held within the state or at least one department that hosts a field camp in a different location outside of the state (Figure 3.27). International field camps are also becoming popular for departments to host with 12 different international destinations recently visited by geoscience students. If a geoscience student did not participate in a field camp, then it is likely that student participated in at least one field course and/or one field experience before graduating (Figure 3.28 and 3.29). Recently, AGI's Directory of Geoscience Departments has been documenting those departments that have a field camp with open or closed enrollment in order to help students find a program to apply to for a summer field camp.

Research experiences were not as common as field experiences, particularly among bachelor's graduates, with 26% completing their degree without participating in a faculty-directed or individual research project (Figure 3.30). Among those students that participated in individual research projects, there are clear differences in how the research was conducted between undergraduate and graduate students. Specifically, graduate students participate in literature-based research and computer-based research more often than undergraduates (Figure 3.31). However, research methods can differ in different geoscience fields. Approximately 91% of master's graduates do participate in individual research projects while working towards their degree, at least 5,183 students published their master's theses in the 2000's (Figure 3.32). For doctoral dissertations, GeoRef reported 10,301 published during the 2000's (Figure 3.33).

Internships should be considered key experiences necessary for anyone considering entering the geoscience workforce because they are recognized by many industries as a good recruitment tool for future employees. Internships can also help students develop many of the needed professional skills, such as effective communication, collaboration, and networking, which may not be as easily gained through their education program. Therefore it is surprising that in 2015, 64% of bachelor's graduates, 40% of master's graduates, and 57% of doctoral graduates did not participate in an internship while working towards their degree. This trend has been seen over the past few years, so AGI's Geoscience Student Exit Survey added questions about the search for internship opportunities. Largely, master's graduates recognize the importance of these opportunities because they tend to submit multiple applications and use any resources available to them to find these opportunities, whereas internships do not appear to be a priority among bachelor's graduates.

Figure 3.25: Field Camp Attendance, 1998-2015

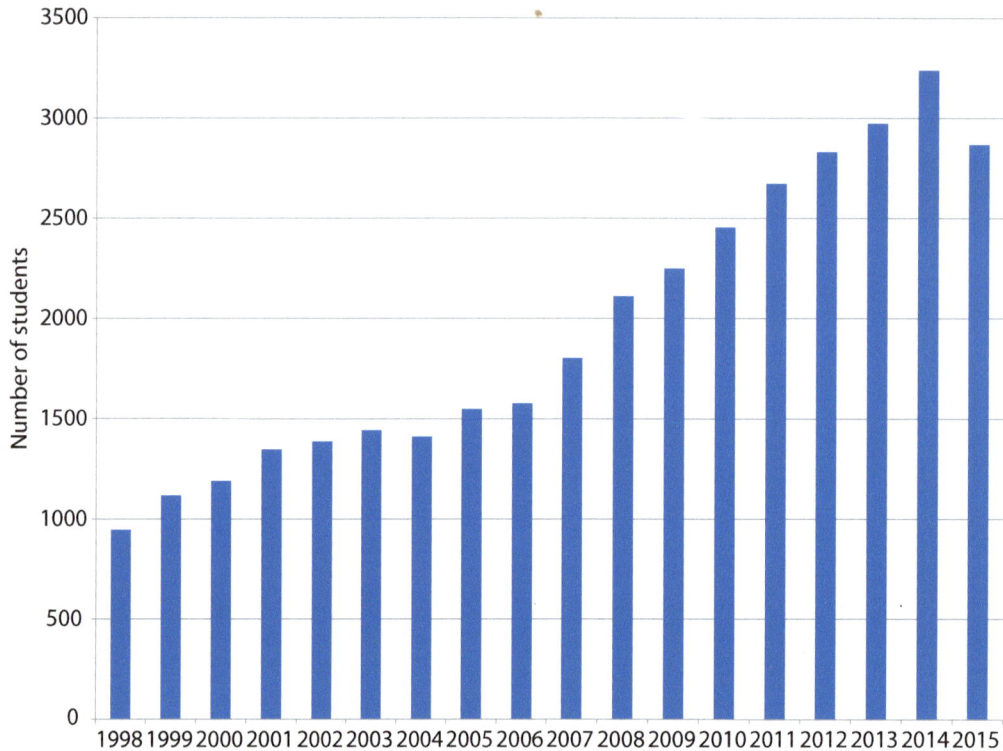

AGI Geoscience Workforce Program; Data provided by Dr. Penelope Morton, UMN-Duluth

Figure 3.26: Geoscience Graduates that Have Participated in a Field Camp, 2015

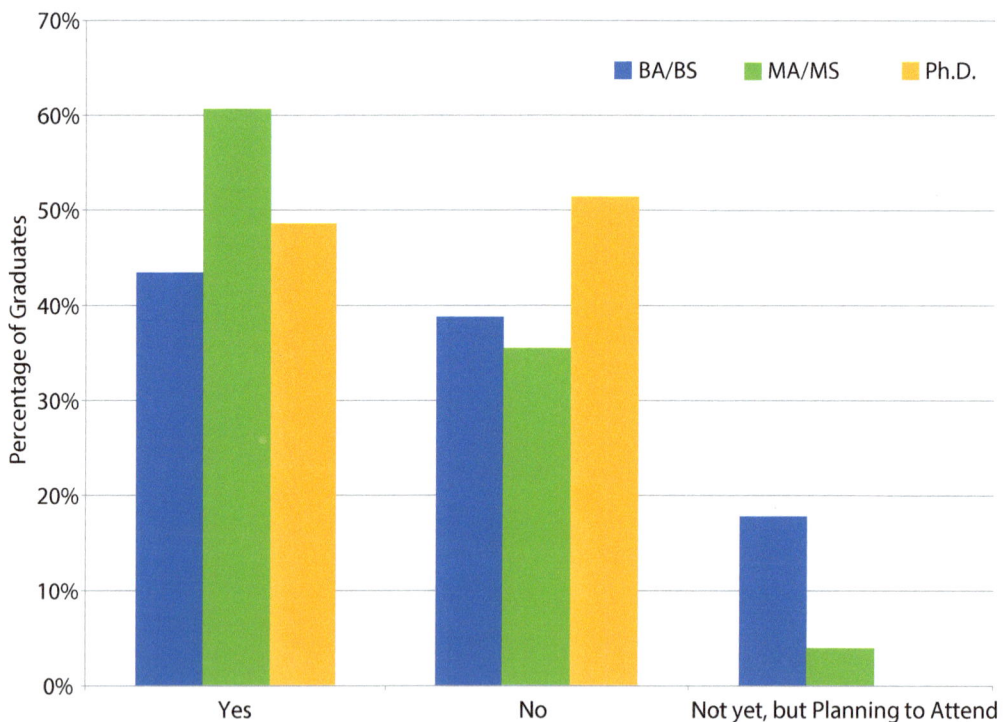

AGI Geoscience Workforce Program; Data derived from AGI's Geoscience Student Exit Survey 2015

Figure 3.27: Locations of Geoscience Field Camps by State, 2015

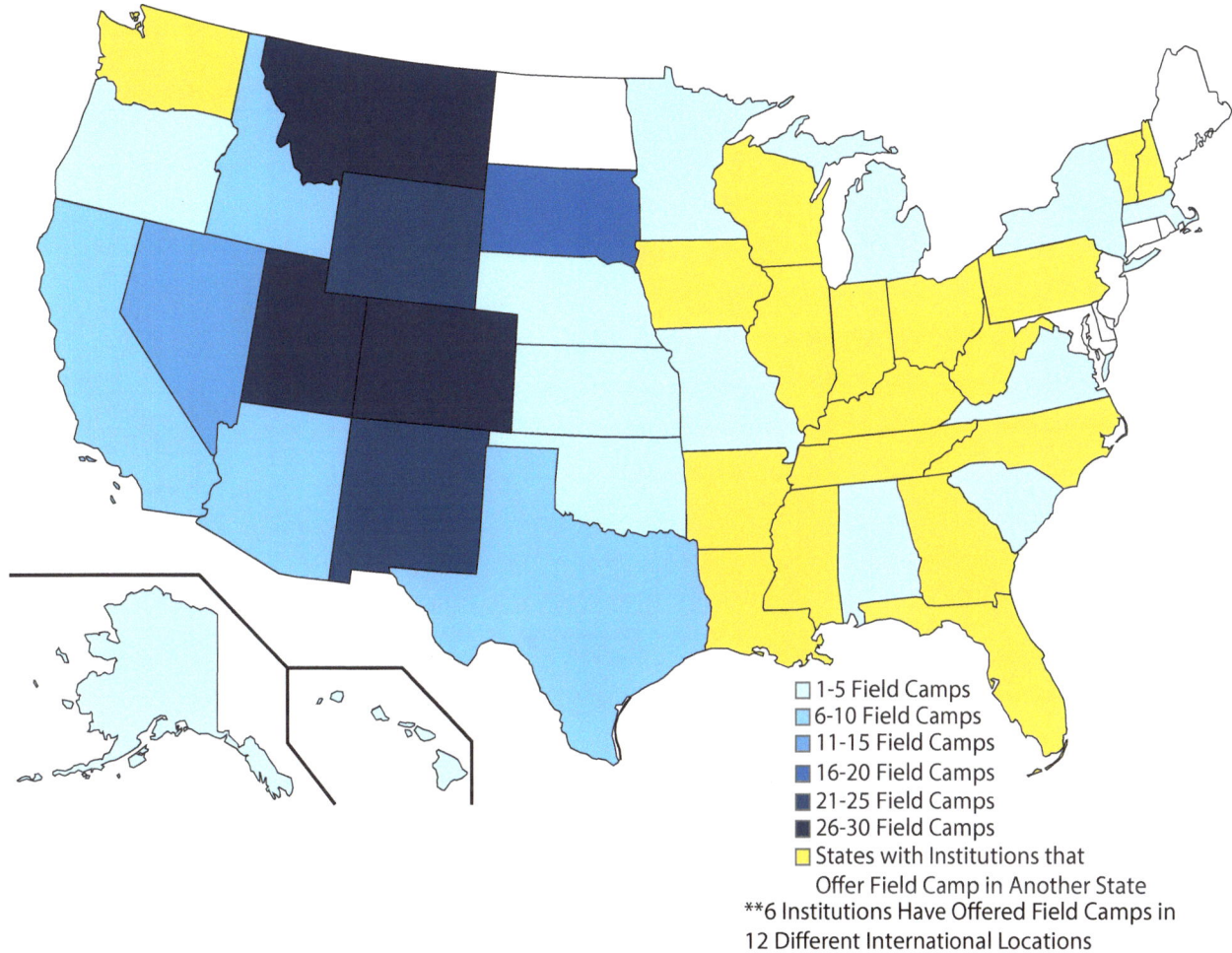

Legend:
- 1-5 Field Camps
- 6-10 Field Camps
- 11-15 Field Camps
- 16-20 Field Camps
- 21-25 Field Camps
- 26-30 Field Camps
- States with Institutions that Offer Field Camp in Another State

**6 Institutions Have Offered Field Camps in 12 Different International Locations

Data provided by Dr. Penelope Morton, UMN-Duluth

Table 3.8: U.S. Universities Hosting Geoscience Field Camps, 2015

University Hosting Field Camp	State	University Hosting Field Camp	State	University Hosting Field Camp	State
University of Alaska-Anchorage	AK	Fort Hays State University	KS	Oklahoma State University	OK
University of Alaska-Fairbanks	AK	University of Kansas	KS	University of Oklahoma	OK
Auburn University	AL	Wichita State University	KS	Oregon State University	OR
University of Alabama	AL	University of Kentucky	KY	Southern Oregon University	OR
University of South Alabama	AL	Louisiana State University	LA	University of Oregon	OR
University of Arkansas	AR	University of Louisiana at Lafayette	LA	Lehigh Universtiy	PA
Arizona State University	AZ	Albion College	MI	Penn State University	PA
Northern Arizona University	AZ	Michigan Technological University	MI	Clemson University	SC
California State University-Chico	CA	University of Michigan	MI	Black Hills Natural Sciences Field Station	SD
California State University-Fullerton	CA	Western Michigan University	MI	University of Memphis	TN
California State University-Longbeach	CA	University of Minnesota-Duluth	MN	Baylor University	TX
California State University-Northridge	CA	University of Minnesota-Twin Cities	MN	Stephen F. Austin State University	TX
Humboldt State University	CA	Missouri University of Science & Technology	MO	Sul Ross State University	TX
San Jose State University	CA	University of Missouri -Columbia	MO	University of Houston	TX
University of California-Santa Barbara	CA	University of Missouri-Kansas City	MO	University of Texas-Arlington	TX
University of California-Santa Cruz	CA	Montana State University	MT	University of Texas-Austin	TX
Adams State College	CO	Montana Tech of the University of Montana	MT	University of Texas-Dallas	TX
Colorado Mesa University	CO	Appalachian State University	NC	University of Texas-El Paso	TX
Colorado School of Mines	CO	University of North Carolina	NC	Brigham Young University	UT
Colorado State University	CO	University of North Carolina-Wilmington	NC	Southern Utah University	UT
Fort Lewis College	CO	University of Nebraska	NE	University of Utah	UT
Western State College of Colorado	CO	Dartmouth College	NH	Utah State University	UT
Florida State University	FL	New Mexico Institute of Mining and Technology	NM	Wasatch Uinta Field Camp	UT
Miami University	FL	University of New Mexico	NM	Weber State University	UT
University of Florida	FL	University of Nevada-Las Vegas	NV	George Mason University	VA
University of Florida	FL	University of Nevada-Reno	NV	James Madison University	VA
Georgia State University	GA	Colgate University	NY	School of International Training	VT
University of Georgia	GA	Cornell University	NY	Central Washington University	WA
Iowa State University	IA	State University of New York-Cortland	NY	Eastern Washington University	WA
Boise State University	ID	State University of New York-Oswego	NY	University of Washington	WA
Brigham Young University-Idaho	ID	University at Buffalo	NY	Washington State University	WA
Idaho State University	ID	Bowling Green State University	OH	Western Washington University	WA
University of Idaho	ID	Kent State University	OH	Beloit College	WI
Illinois State University	IL	Ohio State University	OH	Northland College	WI
Northern Illinois University	IL	Ohio University	OH	University of Wisconsin-Eau Claire	WI
Southern Illinois University	IL	University of Akron	OH	University of Wisconsin-Oshkosh	WI
Western Illinois University	IL			Concord University	WV
Wheaton College	IL			University of West Virginia	WV
Ball State University	IN			University of Wyoming	WY
Indiana University	IN				

Data provided by Dr. Penelope Morton, UMN-Duluth

Figure 3.28: Geoscience Graduates with One or More Field Experiences, 2015

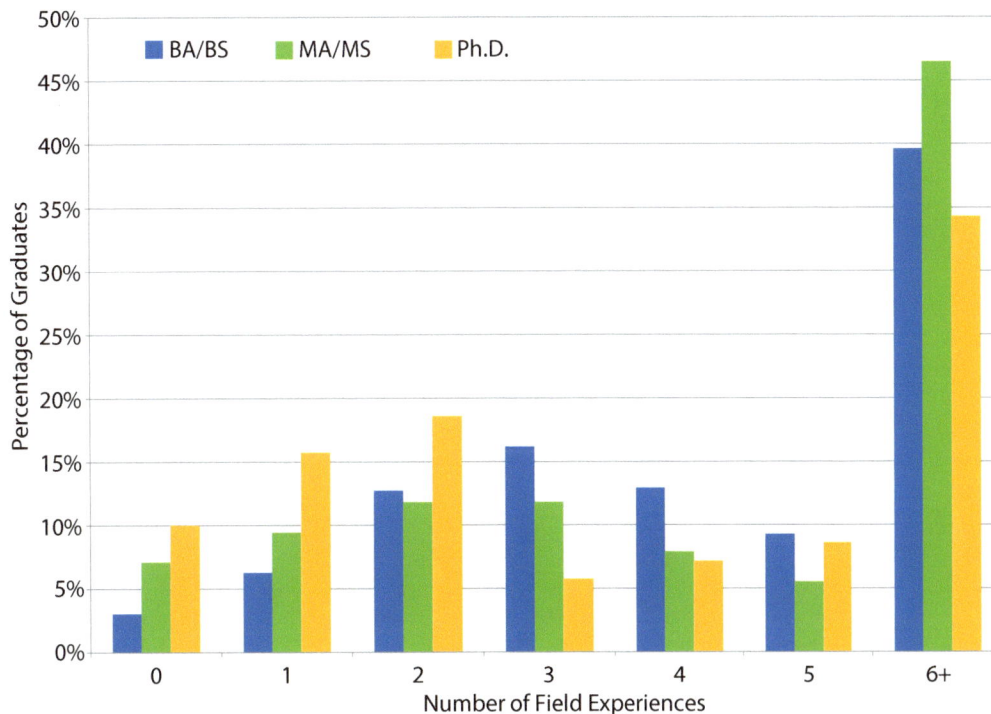

AGI Geoscience Workforce Program, Data derived from AGI's Geoscience Student Exit Survey 2015

Figure 3.29: Geoscience Graduates that Have Taken One or More Field Courses, 2015

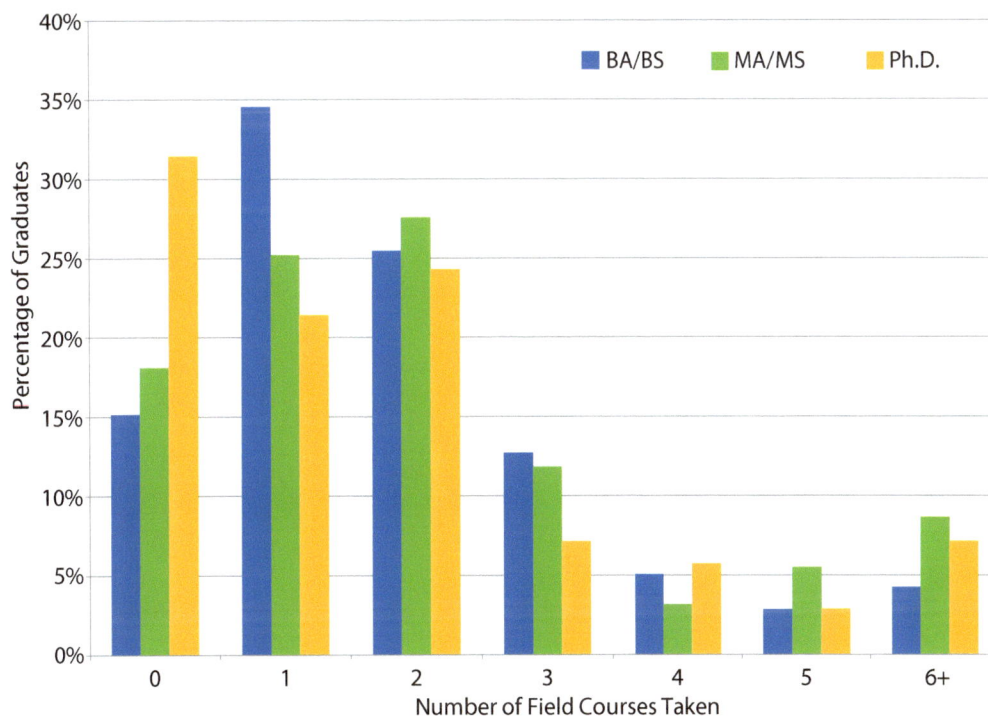

AGI Geoscience Workforce Program; Data derived from AGI's Geoscience Student Exit Survey 2015

Figure 3.30: Geoscience Graduates with One or More Research Experiences, 2015

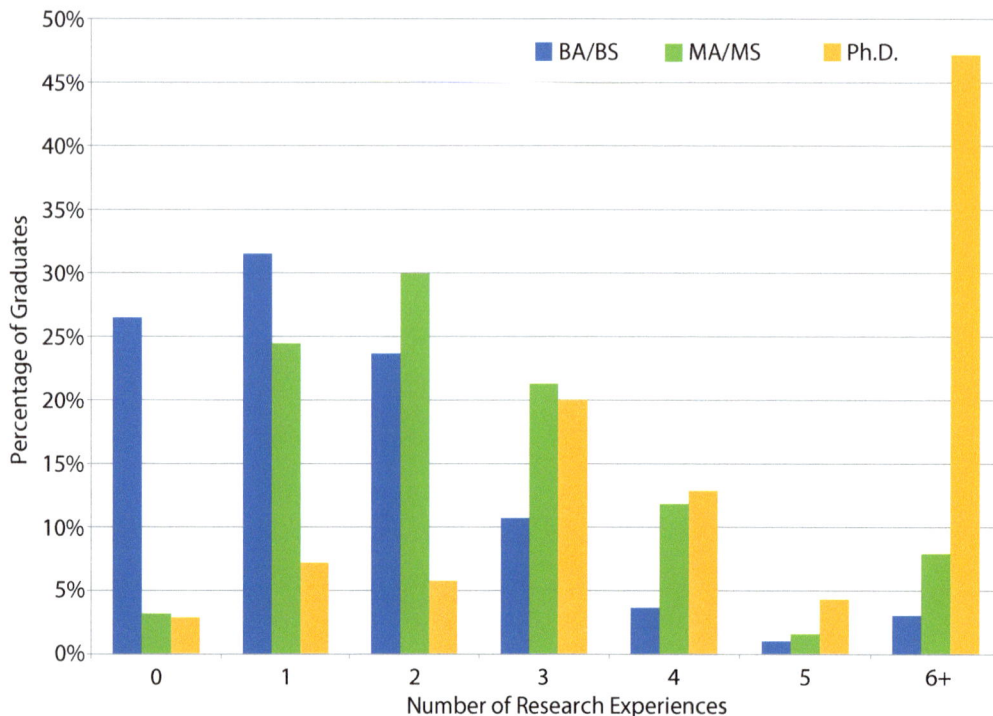

AGI Geoscience Workforce Program; Data derived from AGI's Geoscience Student Exit Survey 2015

Figure 3.31: Research Methods Utilized by Geoscience Graduates for Their Individual Research Projects, 2015

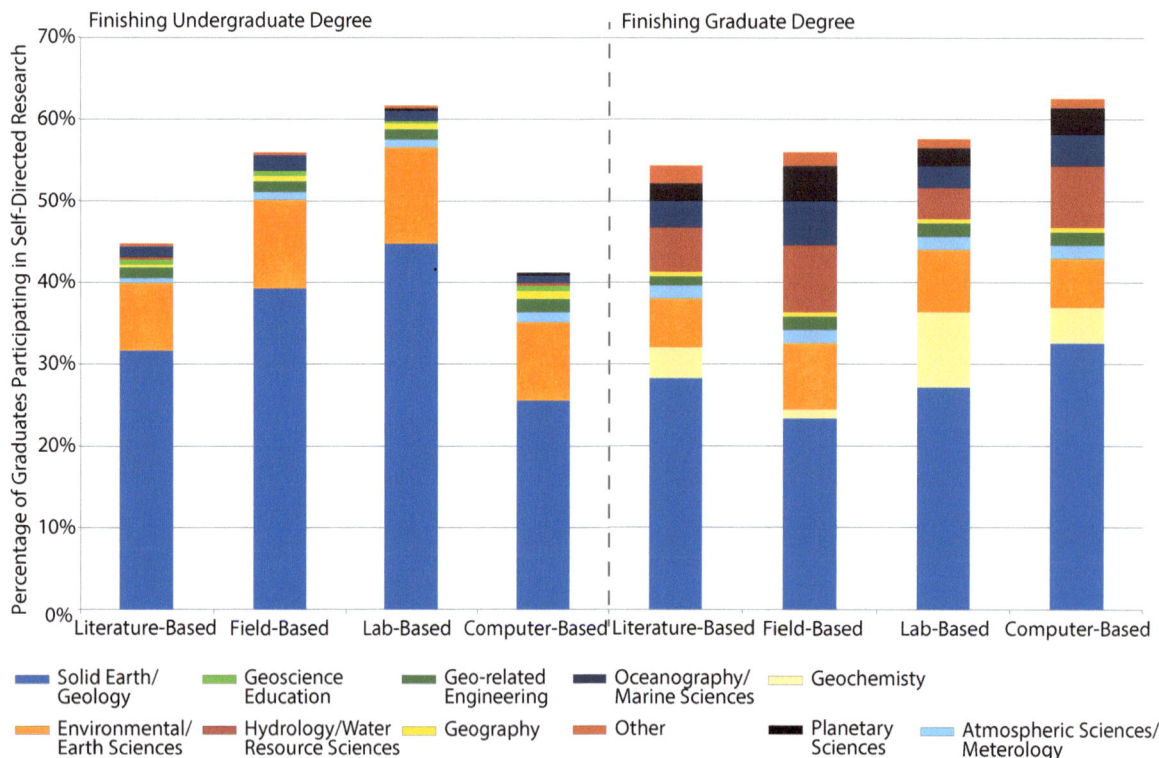

AGI Geoscience Workforce Program; Data derived from AGI's Geoscience Student Exit Survey 2015

Figure 3.32: Trends in Geoscience Master's Thesis Topics

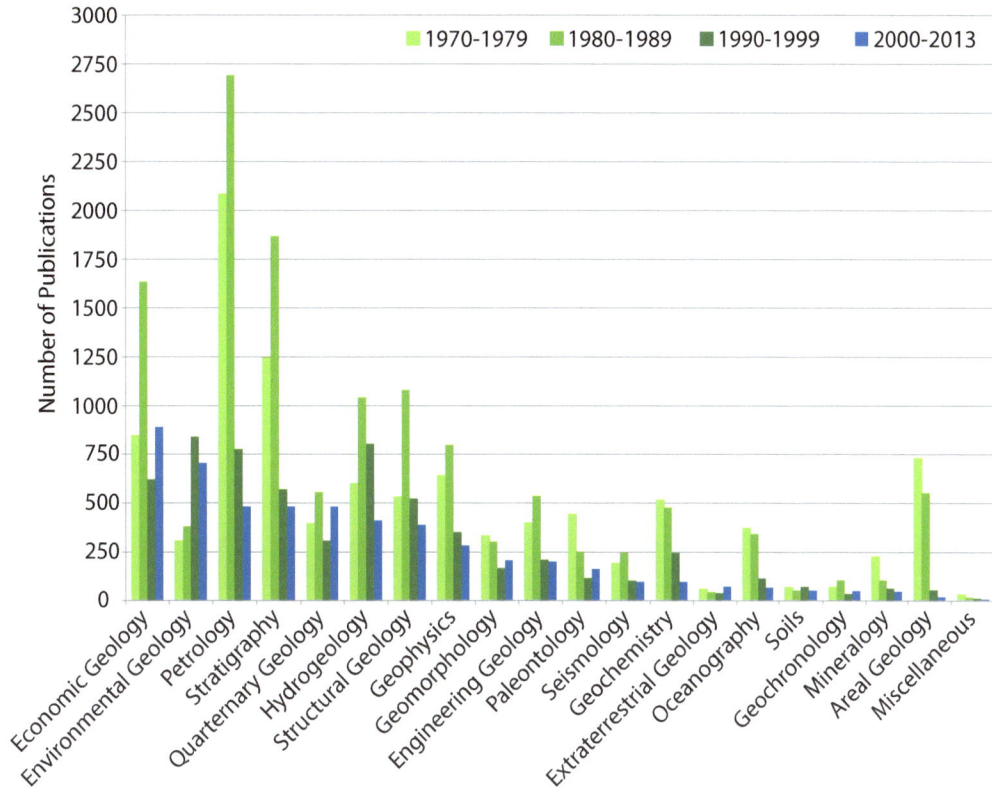

AGI Geoscience Workforce Program; Data derived from AGI's GeoRef database

Figure 3.33: Trends in Geoscience Doctoral Dissertation Topics

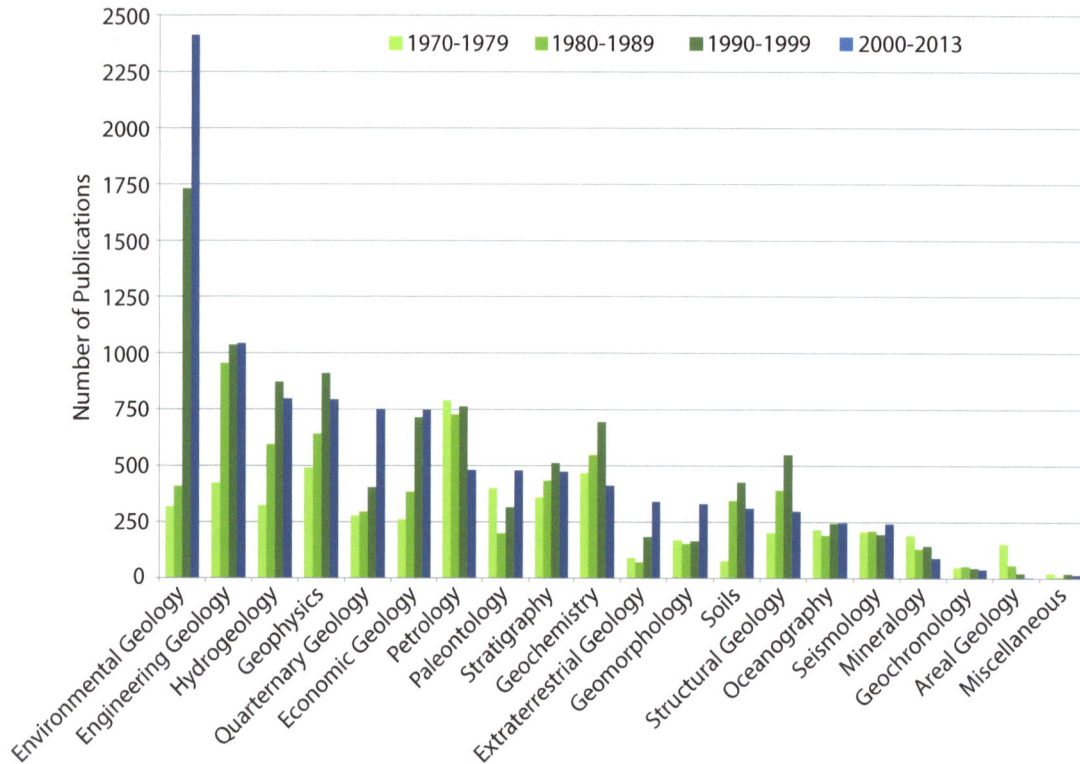

AGI Geoscience Workforce Program; Data derived from AGI's GeoRef database

Table 3.9: Top Five Geoscience Master's Theses Topics

1970-1979	1980-1989	1990-1999	2000-2013
Petrology	Petrology	Environmental Geology	Economic Geology
Stratigraphy	Stratigraphy	Hydrogeology	Environmental Geology
Economic Geology	Economic Geology	Petrology	Petrology
Areal Geology	Structural Geology	Economic Geology	Stratigraphy
Geophysics	Hydrogeology	Stratigraphy	Quaternary Geology

AGI Geoscience Workforce Program; Data derived from AGI's GeoRef database

Table 3.10: Top Five Geoscience Doctoral Dissertation Topics

1970-1979	1980-1989	1990-1999	2000-2013
Petrology	Engineering Geology	Environmental Geology	Environmental Geology
Geophysics	Petrology	Engineering Geology	Engineering Geology
Geochemistry	Geophysics	Geophysics	Hydrogeology
Engineering Geology	Hydrogeology	Hydrogeology	Geophysics
Paleontology	Geochemistry	Petrology	Quaternary Geology

AGI Geoscience Workforce Program; Data derived from AGI's GeoRef database

Figure 3.34: Number of Internships Held by Geoscience Graduates, 2015

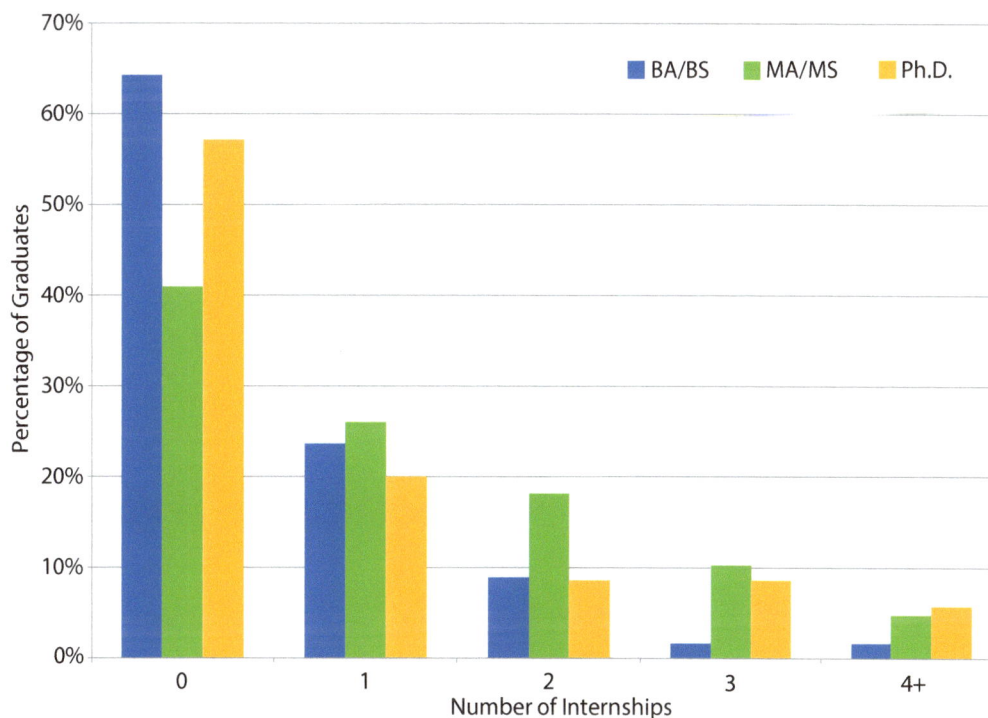

AGI Geoscience Workforce Program; Data derived from AGI's Geoscience Student Exit Survey 2015

Student Plans for Graduate School

Among recent geoscience graduates in 2015, 38% of bachelor's graduates and 20% of master's graduates planned to immediately return to school for a graduate degree (Figure 3.35). Concern has been raised that many of the geoscience graduates programs have reached their capacity and cannot accept many of the applicants to their programs. It appears that students have been made aware of these concerns because there has been a decrease in the percentages of bachelor's and master's graduates planning to immediately enroll in a graduate program.

Figure 3.35: Geoscience Students Planning to Attend Graduate School Immediately After Graduation, 2015

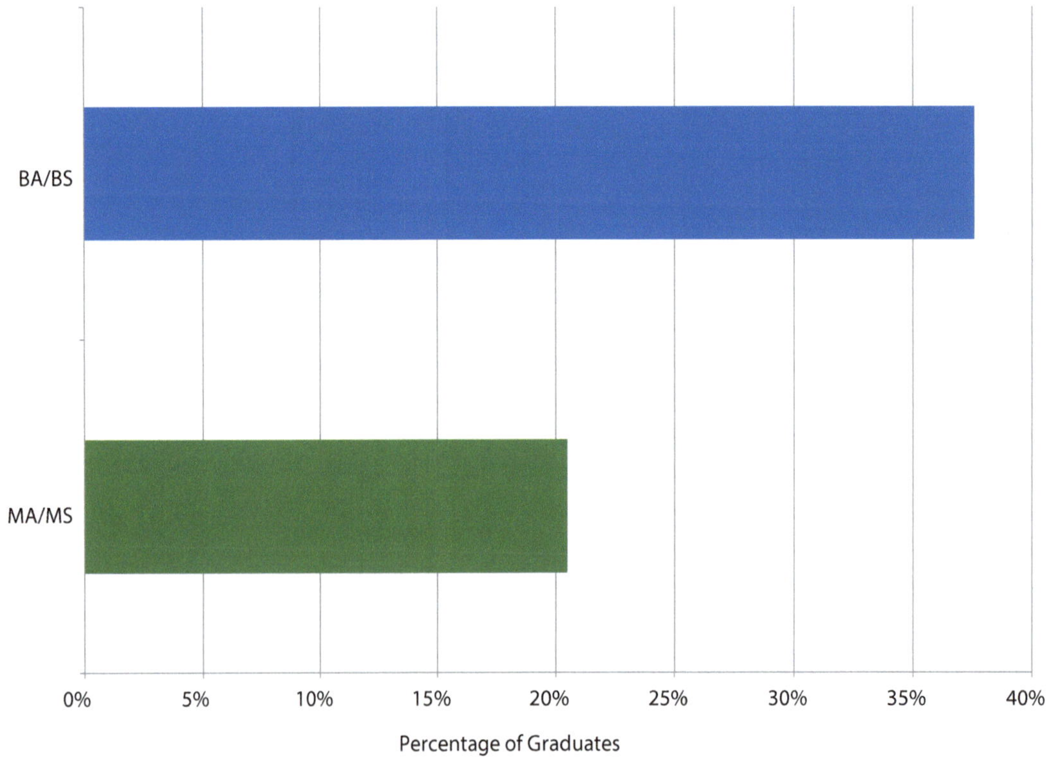

AGI Geoscience Workforce Program; Data derived from AGI's Geoscience Student Exit Survey 2015

Figure 3.36: Geoscience Undergraduate Students Planning to Pursue a Graduate Degree, 2015

Students graduating with an undergraduate degree

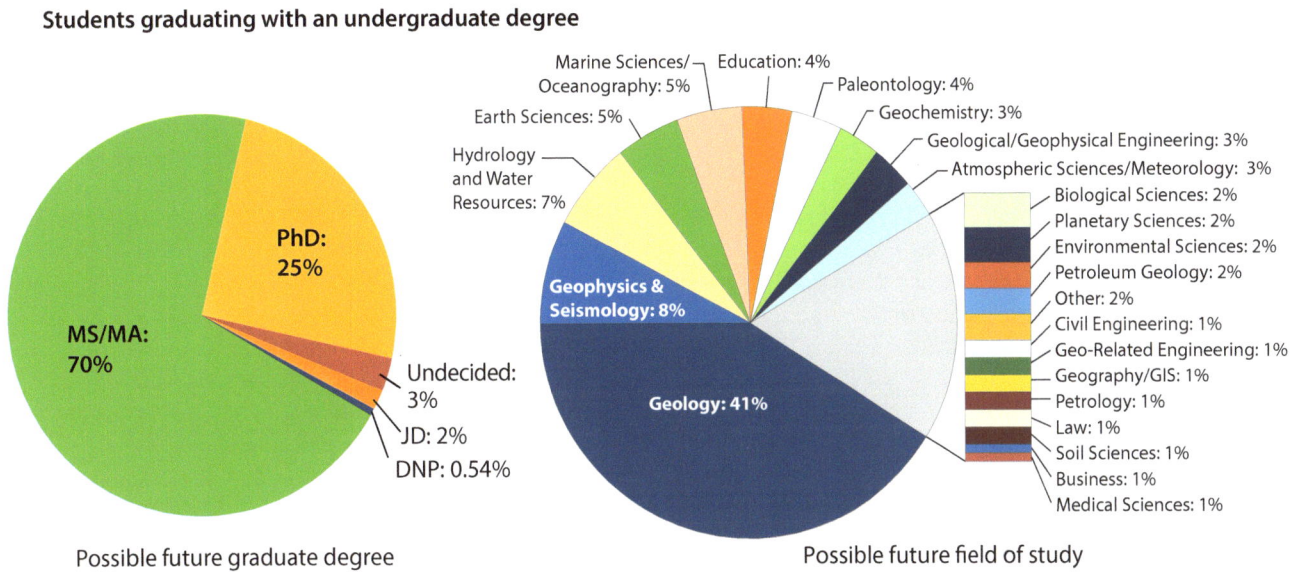

Possible future graduate degree

Possible future field of study

AGI Geoscience Workforce Program; Data derived from AGI's Geoscience Student Exit Survey 2015

Figure 3.37: Geoscience Graduate Students Planning to Pursue Another Graduate Degree, 2015

Students graduating with a graduate degree

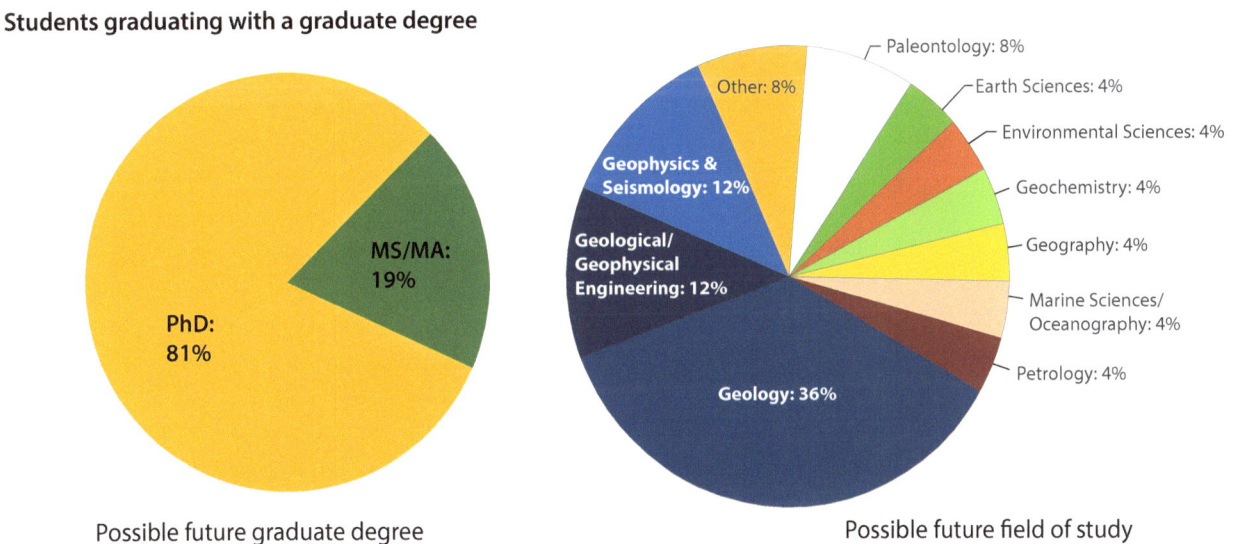

Possible future graduate degree

Possible future field of study

AGI Geoscience Workforce Program; Data derived from AGI's Geoscience Student Exit Survey 2015

Funding of the Geosciences at the University Level

The percentage of federal research funding applied to the geosciences had decreased from a high of 11% in 1996 to a low of 5% in 2010, but it is holding steady around 7% through 2015 (Figure 3.38). However, the overall amount of federal research funds awarded to universities has steadily risen, which highlights the overall increases in total federal research funding since the 1970's (Figure 3.39). The majority of the research funds for geoscience given to universities overwhelmingly came from the National Science Foundation (NSF) reaching $5.96 billion in 2013 (Figure 3.44).

At NSF, the funding rate of geoscience proposals has been on a downward trend since 2009 (Figure 3.45). This downward trend can also be seen for the funding rates within the different geoscience divisions (Figure 3.46). The Atmospheric and Geospace Division reached a high funding rate of 50% in 2012, but it has decreased since

then. In 2009 and 2010, the United States Government introduced the American Recovery and Reinvestment Act (ARRA), and the money from this stimulus program given to NSF allowed for a higher percentage of proposals to receive funding. As a result, the NSF Geoscience Directorate awarded 804 ARRA proposals, which inflated the overall funding rate and the funding rates within the divisions. However the overall funding rate in 2015 was lower than the fairly steady rate seen from 2006-2008 and the number of proposals submitted to the Geoscience Directorate continued to rise beyond 2009 reaching a high from 2013-2015 of around 6,000 proposals, which lends to speculation that the ARRA funding may have lead to unrealistic expectations on the part of the proposal writers on their chances for funding. While the funding rate has been decreasing, the median award size increased to $133,201 in 2013 and has been holding steady at $129,000 for 2014 and 2015 (Figure 3.47).

Figure 3.38: Percentage of Total Federal Research Funding Applied to the Geosciences

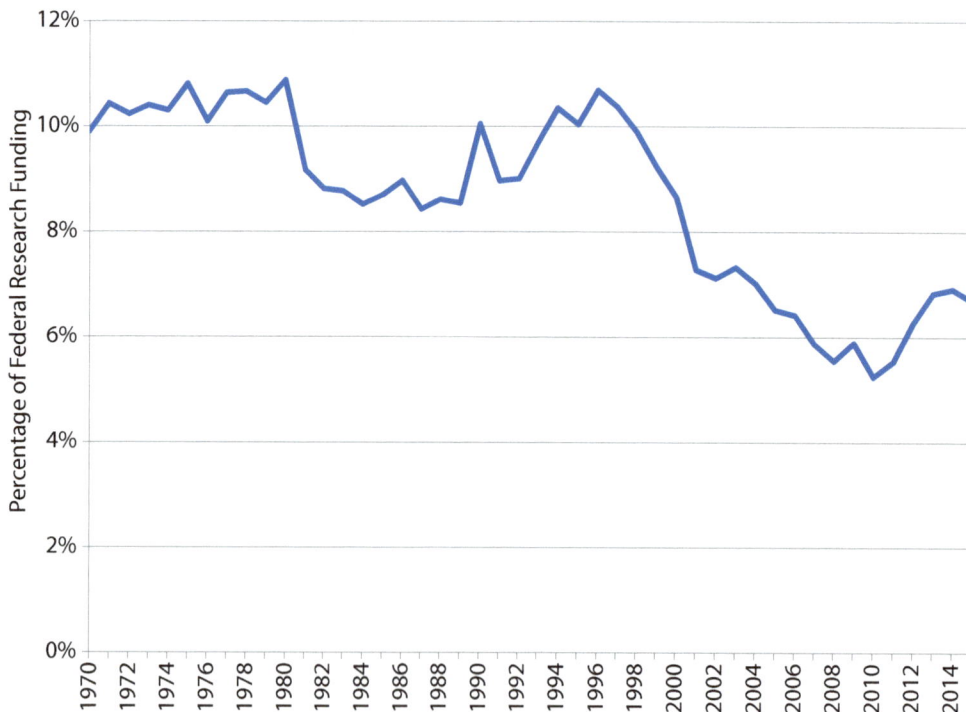

AGI Geoscience Workforce Program; Data derived from NSF/SRS Survey of Federal Funds for Research & Development

Figure 3.39: Percentage of University Geoscience Research Funding per Subdiscipline from Selected Federal Agencies

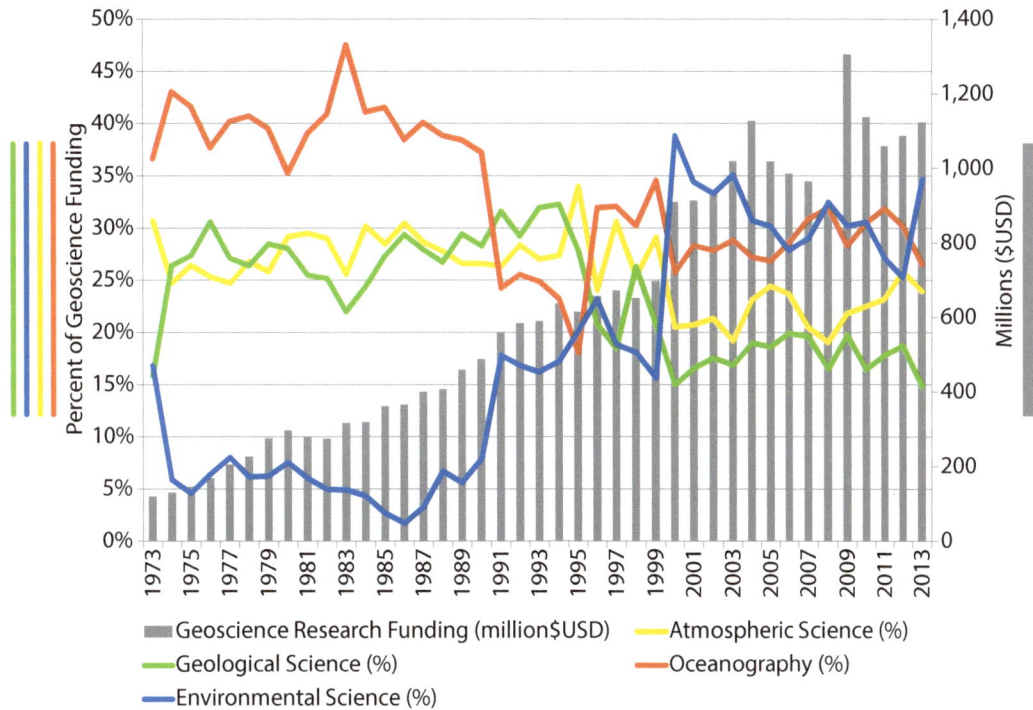

AGI Geoscience Workforce Program; Data derived from NSF/SRS Survey of Federal Funds for Research & Development

Figure 3.40: Average Annual University Geoscience Research Funding by the Department of Agriculture

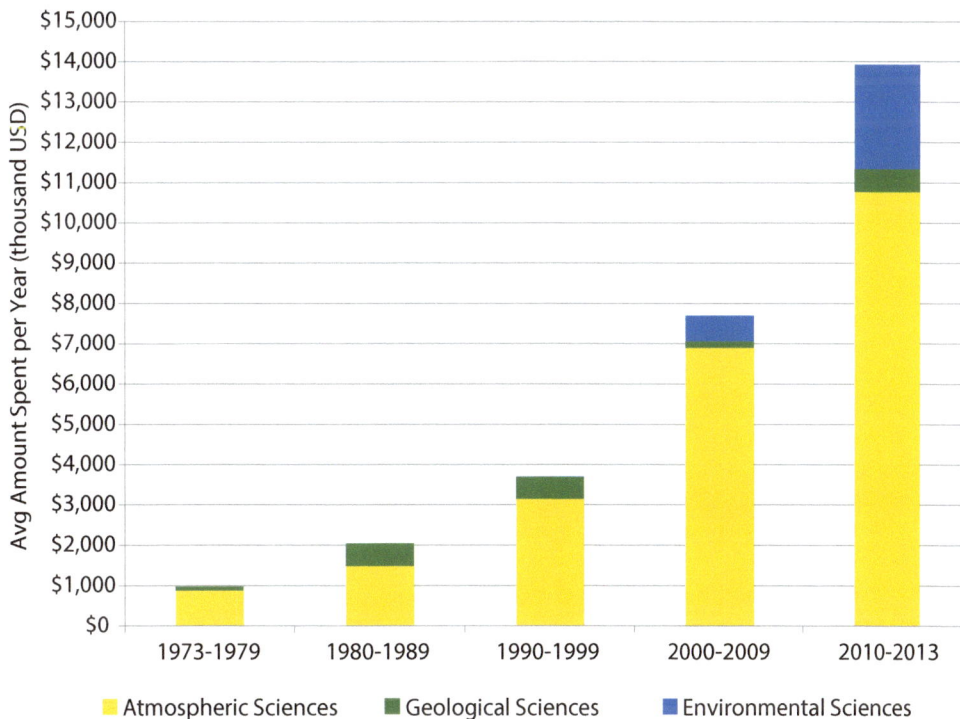

AGI Geoscience Workforce Program; Data derived from NSF/SRS Survey of Federal Funds for Research & Development

Figure 3.41: Average Annual Geoscience Research Funding by the Department of Defense

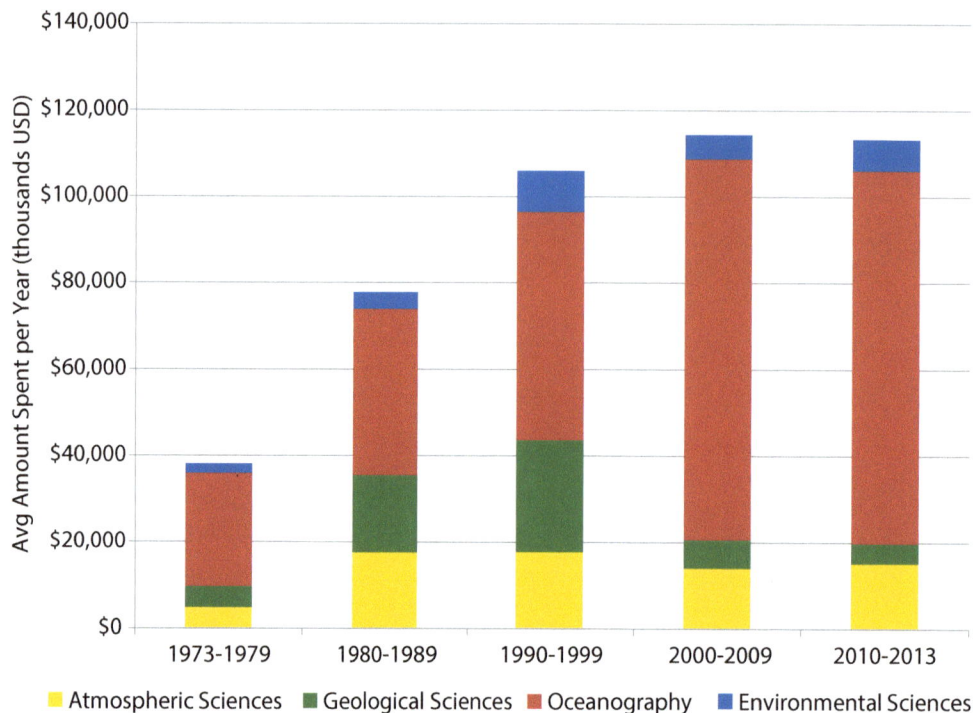

AGI Geoscience Workforce Program; Data derived from NSF/SRS Survey of Feceral Funds for Research & Development

Figure 3.42: Average Annual University Geoscience Research Funding by the Department of Energy

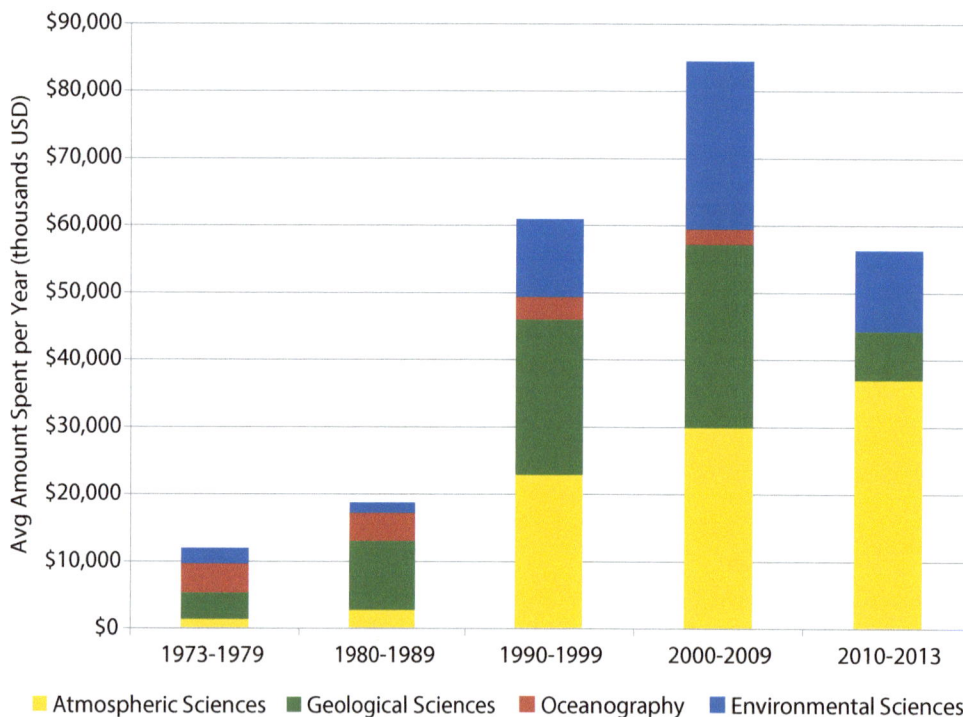

AGI Geoscience Workforce Program; Data derived from NSF/SRS Survey of Federal Funds for Research & Development

Figure 3.43: Average Annual University Geoscience Research Funding by NASA

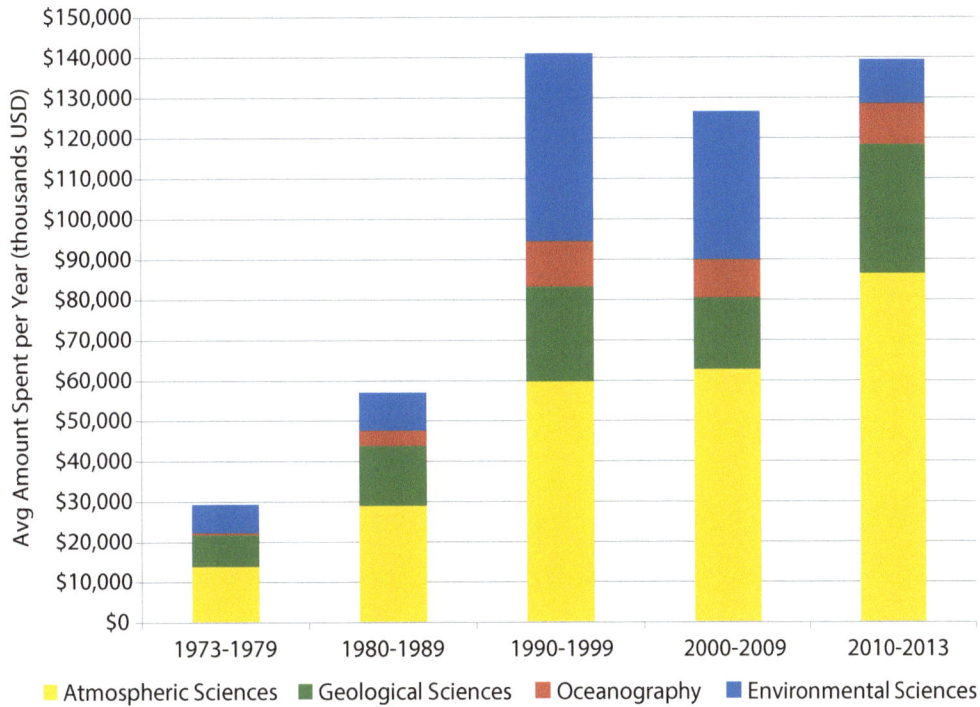

AGI Geoscience Workforce Program; Data derived from NSF/SRS Survey of Federal Funds for Research & Development

Figure 3.44: Average Annual University Geoscience Research Funding by NSF

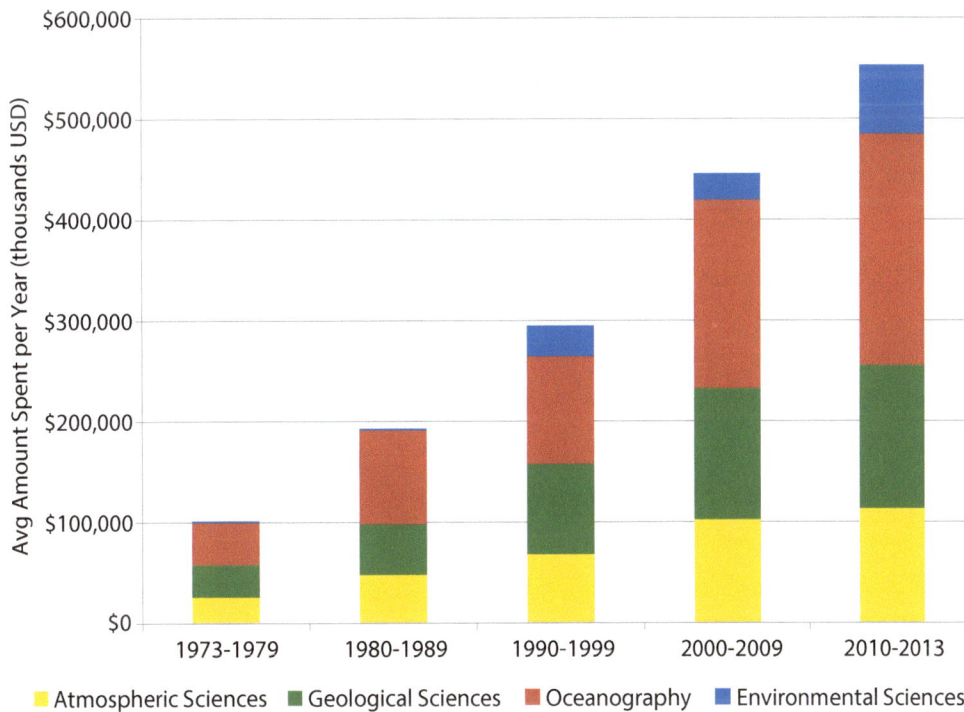

AGI Geoscience Workforce Program; Data derived from NSF/SRS Survey of Federal Funds for Research & Development

Figure 3.45: Funding of Geoscience Proposals at NSF

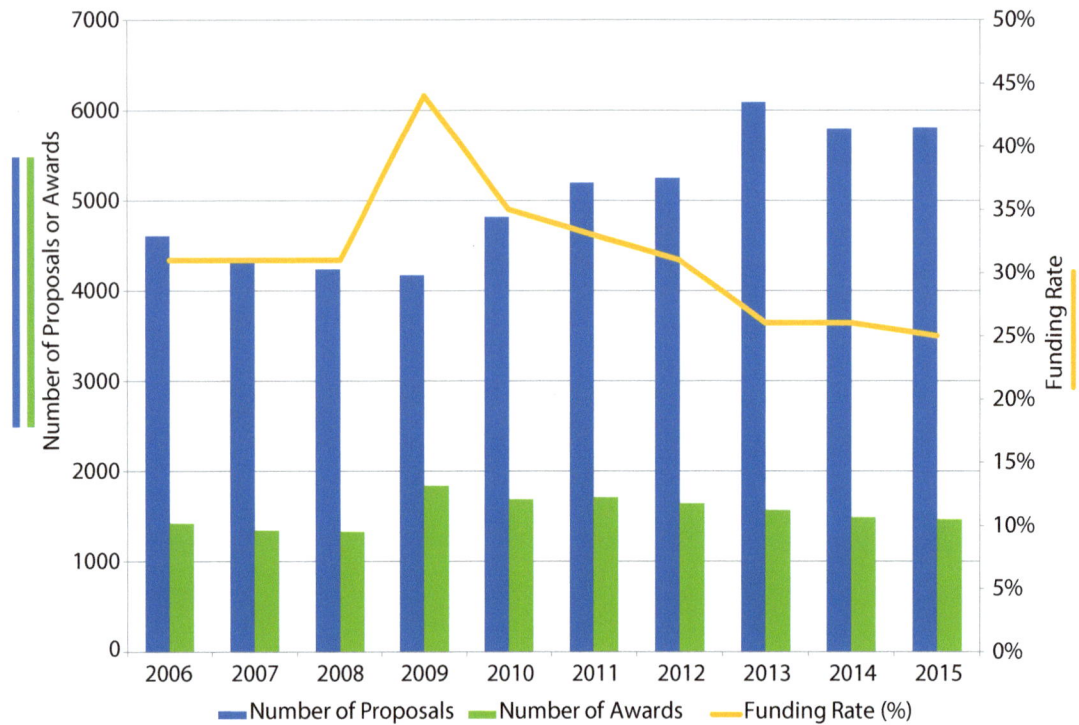

AGI Geoscience Workforce Program; Data derived from NSF's BIIS Funding Trends database

Figure 3.46: Funding Rates of Geoscience Proposals at NSF by GEO Division

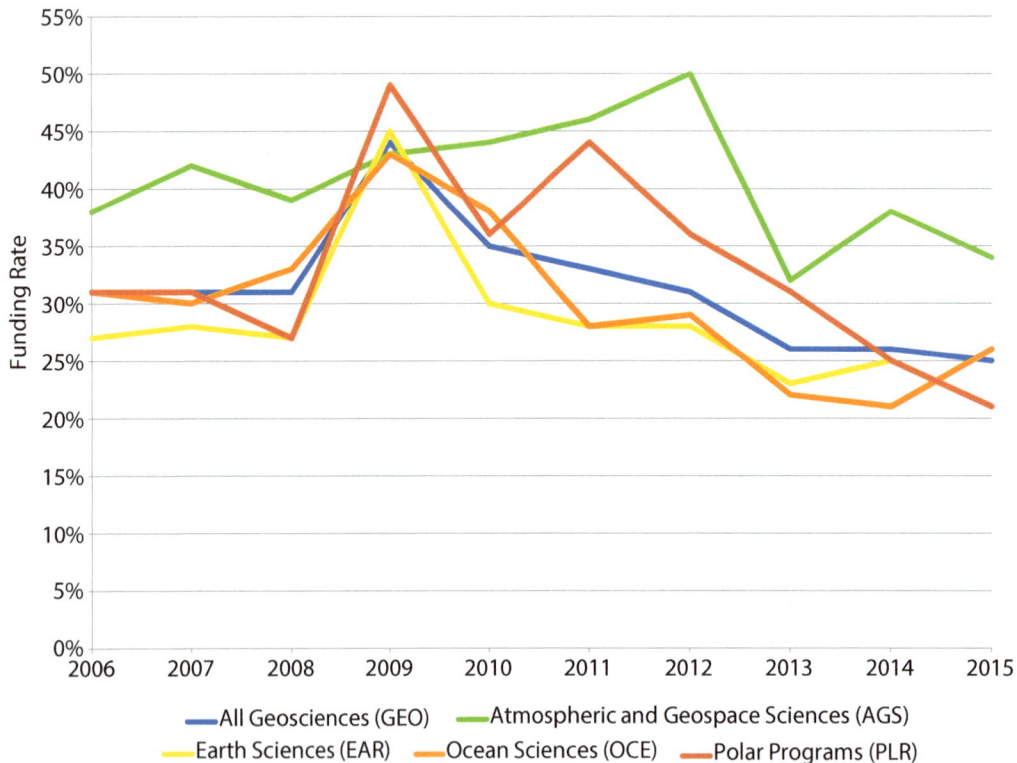

AGI Geoscience Workforce Program; Data derived from NSF's BIIS Funding Trends database

Figure 3.47: Median Annual Size of Geoscience Awards at NSF by GEO Division

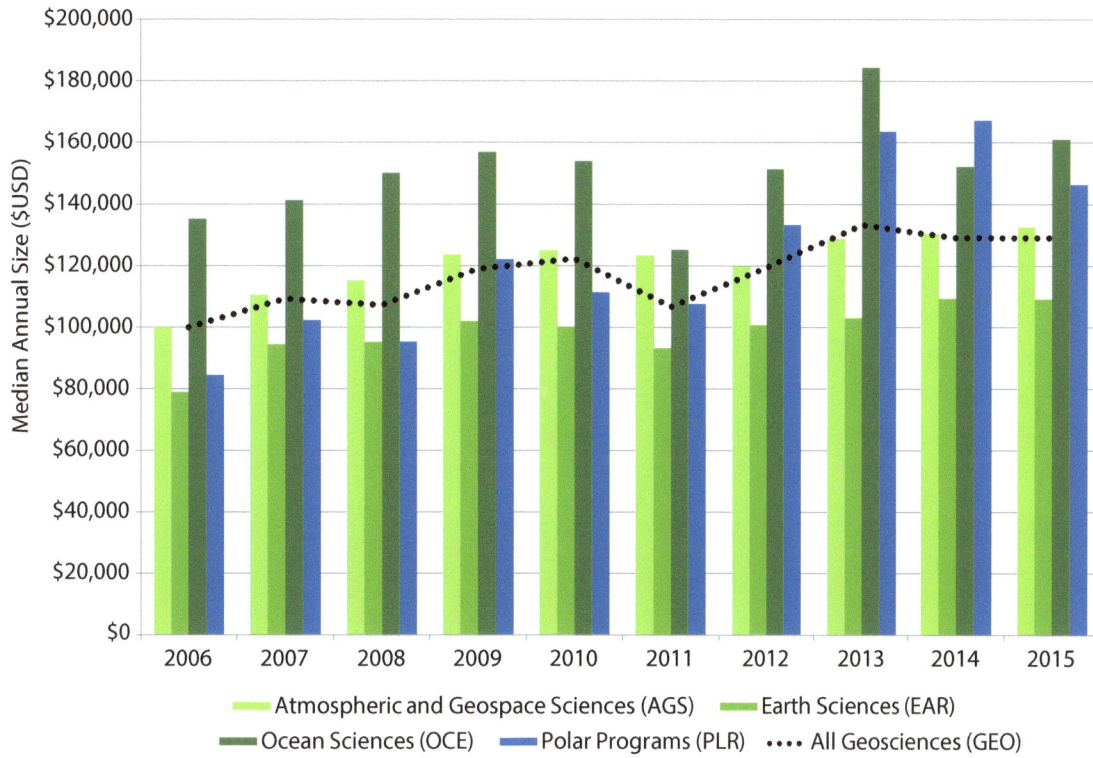

AGI Geoscience Workforce Program; Data derived from NSF's BISS Funding Trends database

Table 3.11: Top 10 Universities Receiving NSF Atmospheric and Geospace Science (AGS) Awards Annually, 2006-2015 (Millions $USD)

Institution	State	2006	2007	2008	2009	2010	2011	2012	2013	2014	2015
Colorado State University	CO	$3.3	$8.3	$8.3	$8.5	$9.2	$7.5	$8.8	$7.6	$8.7	$11.5
University of Colorado - Boulder	CO	$3.0	$4.9	$4.5	$7.3	$4.0	$5.6	$4.7	$4.1	$3.1	$5.1
University of Illinois - Urbana-Champaign	IL	$4.4			$4.5				$2.7		$4.3
Massachusetts Institute of Technology	MA	$3.9		$2.7	$4.6	$4.3	$4.1	$4.2	$2.7	$4.0	$4.2
Carnegie-Mellon University	PA										$2.7
University of Oklahoma	OK	$3.2	$3.4	$3.6					$4.9		$2.6
University of Michigan	MI					$5.1		$2.5	$3.1		$2.6
University of Wyoming	WY						$3.3	$3.6	$4.9		$2.5
University of California - Los Angeles	CA	$3.0		$3.0			$4.3	$4.6		$6.0	$2.3
Pennsylvania State University - University Park	PA		$3.4		$4.4		$3.1	$2.7		$3.1	$2.3
University of Washington	WA	$5.4	$3.4	$3.3	$4.8		$4.7	$5.0	$4.0	$4.4	
University of California - Berkeley	CA									$2.8	
Johns Hopkins University	MD			$2.9	$5.0					$2.6	
Columbia University	NY	$3.4				$3.5	$2.8	$2.7	$4.7		
University of Miami	FL			$2.7					$2.70		
Colorado School of Mines	CO							$2.3			
Boston University	MA	$5.7	$5.6	$5.7	$5.6	$6.2	$4.6				
Cornell University	NY		$5.4	$6.5		$3.0	$3.0				
New Jersey Institute of Technology	NJ					$6.0					
Oregon State University	OR	$3.2				$5.4					
University of Hawaii	HI					$3.1					
Virginia Polytechnic Institute and State University	VA				$7.6						
George Mason University	VA				$4.3						
Scripps Institute of Oceanographic Research	CA		$5.3								
University of Arizona	AZ		$2.8								

AGI Geoscience Workforce Program; Data derived from NSF's BIIS Funding Trends database

Table 3.12: Top 10 Universities Receiving NSF Earth Science (EAR) Awards Annually, 2006-2015 (Millions $USD)

Institution	State	2006	2007	2008	2009	2010	2011	2012	2013	2014	2015
Columbia University	NY	$2.9		$3.8	$4.4	$3.6	$3.1	$3.5	$8.2	$4.2	$6.7
University of Southern California	CA	$4.7	$4.8		$7.2	$5.8	$5.1	$5.6	$5.1	$4.1	$5.4
University of Colorado - Boulder	CO		$3.9	$3.2	$4.9	$4.0	$5.2	$4.1	$5.0	$5.0	$4.8
University of Minnesota - Twin Cities	MN	$5.5	$8.0	$5.9	$7.5	$5.7	$5.1	$3.6	$5.0	$4.5	$4.3
University of Illinois - Urbana-Champaign	IL					$3.8	$3.3	$3.8	$3.0	$4.0	$3.8
University of California - Berkeley	CA	$2.8									$3.7
University of Chicago	IL		$2.7	$2.7		$3.6	$3.2	$5.2			$3.1
University of Arizona	AZ	$6.6	$6.3	$5.4	$11.6			$3.2	$4.4	$4.2	$3.0
Oregon State University	OR		$4.4								$2.8
University of California - Los Angeles	CA									$4.6	$2.7
University of Wisconsin - Madison	WI					$4.0			$3.0	$3.8	
California Institute of Technology	CA	$4.6	$4.0	$4.0	$5.1				$3.1	$3.2	
Indiana University	IN									$2.9	
Pennsylvania State University - University Park	PA	$2.7		$3.0	$3.9	$4.2			$3.5		
University of Texas - Austin	TX			$2.7		$3.8			$3.0		
Arizona State University	AZ						$3.6	$4.0			
University of California-Davis	CA							$3.7			
Woods Hole Oceanographic Institute	MA							$3.1			
Scripps Institute of Oceanography	CA					$3.1					
Massachusetts Institute of Technology	MA	$3.1	$3.3	$3.1	$4.9		$3.0				
University of Pennsylvania	PA				$4.8						
State University of New York - Stony Brook	NY	$3.7	$3.3	$3.5	$4.5						
Stanford University	CA	$6.8	$6.3								

AGI Geoscience Workforce Program; Data derived from NSF's BIIS Funding Trends database

Table 3.13: Top 10 Universities Receiving NSF Ocean Sciences (OCE) Awards Annually, 2006-2015 (Millions $USD)

Institution	State	2006	2007	2008	2009	2010	2011	2012	2013	2014	2015
Texas A&M University	TX										$48.0
Woods Hole Oceanographic Institution	MA	$58.6	$62.1	$53.6	$74.0	$63.5	$59.6	$50.8	$55.7	$50.2	$44.5
Columbia University	NY	$17.4	$12.4	$22.8	$27.4	$17.9	$20.0	$23.7	$17.5	$18.9	$28.5
Scripps Institute of Oceanographic Research	CA	$28.0	$28.8	$24.4	$36.0	$23.5	$25.7	$31.1	$21.0	$21.7	$22.0
University of Washington	WA	$13.8	$19.1	$15.8	$21.4	$13.6	$14.2	$20.2	$9.5	$11.4	$16.6
University of Hawaii	HI	$17.5	$16.5	$12.3	$13.8	$17.3	$7.6	$12.1	$13.3	$10.9	$14.4
Oregon State University	OR	$13.0	$11.3	$10.6	$12.3	$7.4	$12.7	$9.4	$11.8	$14.5	$10.5
University of Miami	FL	$9.8	$6.7	$6.6	$9.9	$7.1	$6.4	$5.2	$5.6	$4.2	$7.0
University of Southern California	CA						$8.5	$6.1	$9.1	$6.9	$6.4
University of California - Santa Barbara	CA		$3.6				$5.0		$5.4		$5.4
University Rhode Island	RI	$4.5	$5.2	$4.9	$10.1	$6.3	$6.5	$4.7		$5.2	
University of Alaska - Fairbanks	AK				$162.6	$34.0				$4.4	
Oregon Health and Science University	OR		$4.2	$4.0					$4.0		
University of California-Santa Cruz	CA	$2.7						$4.5			
University of Georgia	GA					$4.8					
Georgia Institute of Technology	GA				$4.5						
Massachusetts Institute of Technology	MA			$4.1							
University of Delaware	DE	$2.5									

AGI Geoscience Workforce Program; Data derived from NSF's BIIS Funding Trends database

Figure 3.48: Trends in NSF Atmospheric and Geospace Science Funding Rates by Subject

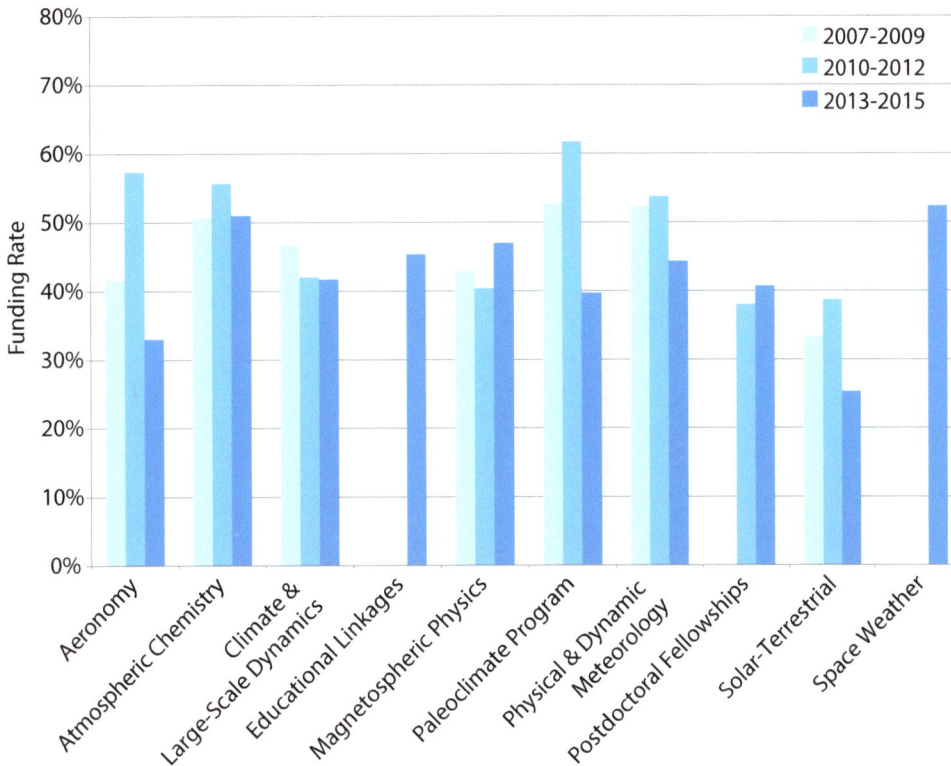

AGI Geoscience Workforce Program; Data derived from NSF's BISS Funding Trends database

Figure 3.49: Trends in NSF Atmospheric and Geoscience Sciences Award Size by Subject

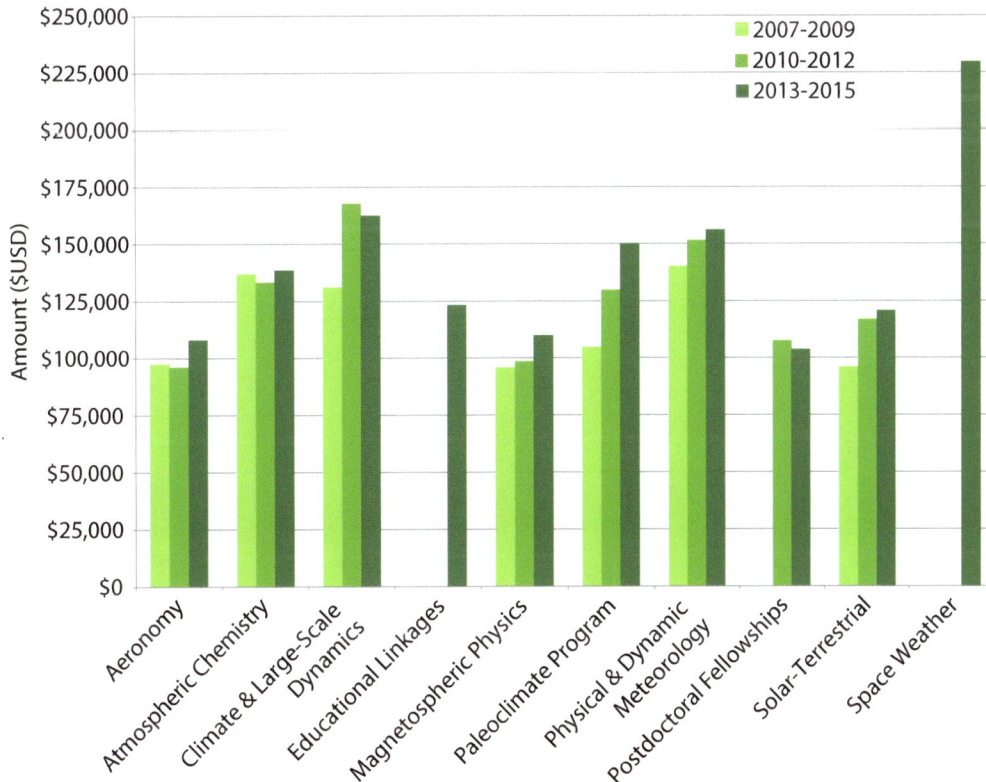

AGI Geoscience Workforce Program; Data derived from NSF's BIIS Funding Trends database

Figure 3.50: Trends in NSF Earth Sciences Funding Rates by Subject

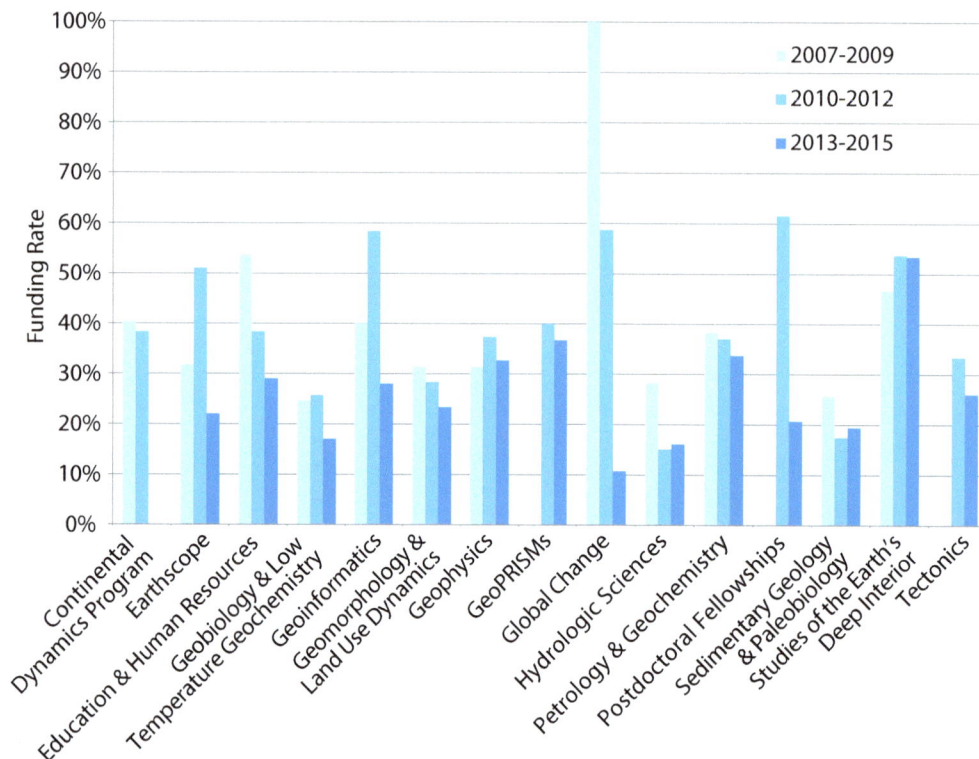

AGI Geoscience Workforce Program; Data derived from NSF's BIIS Funding Trends database

Figure 3.51: Trends in NSF Earth Sciences Award Size by Subject

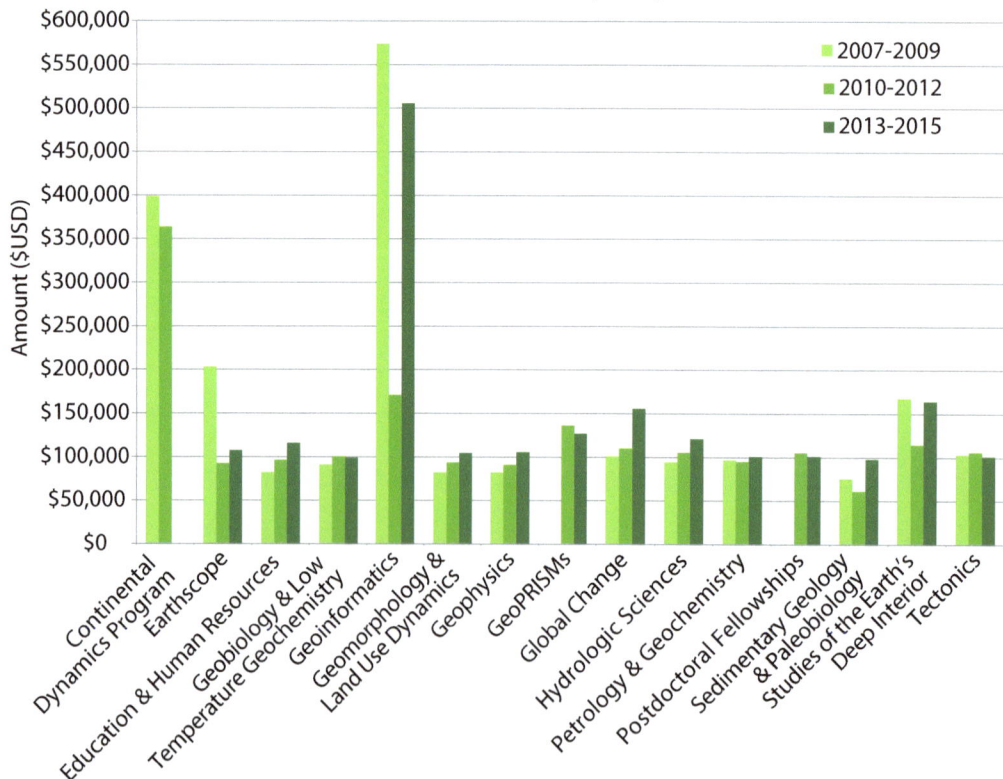

AGI Geoscience Workforce Program; Data derived from NSF's BIIS Funding Trends database

Figure 3.52: Trends in NSF Ocean Sciences Funding Rates by Subject

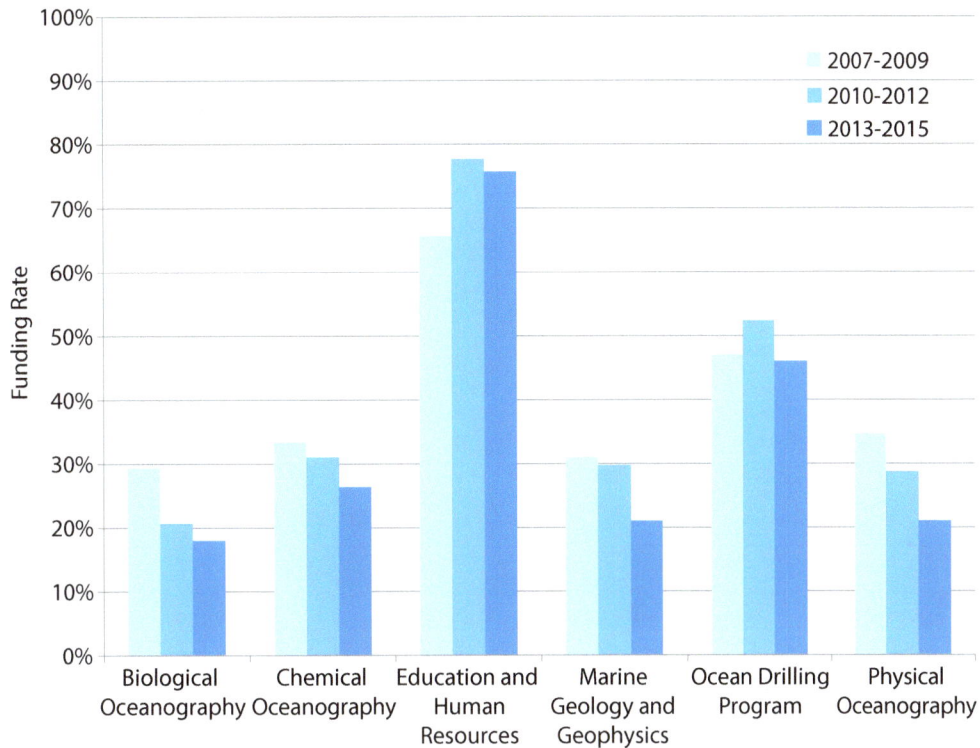

AGI Geoscience Workforce Program; Data derived from NSF's BIIS Funding Trends database

Figure 3.53: Trends in NSF Ocean Sciences Award Size by Subject

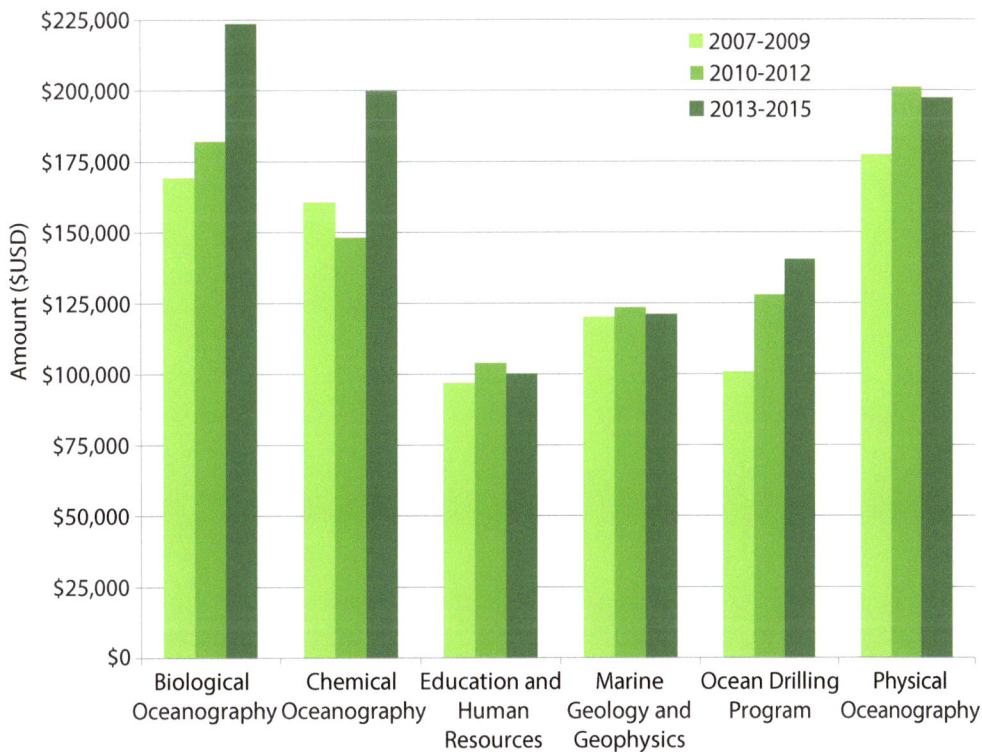

AGI Geoscience Workforce Program; Data derived from NSF's BIIS Funding Trends database

Funding of Geoscience Students

Geoscience students use a variety of funding sources to help pay for their degree programs, including student loans, teaching assistantships, research assistantships, federal grants, and institutional scholarships. It is somewhat surprising that 35% of master's graduates and 14% of doctoral graduates used student loans to help pay for their degree program considering anecdotal discussions assume students in geoscience graduate programs get tuition and fees covered by the institution (Figure 3.53).

NSF's Graduate Research Fellowships are a prestigious and well-funded award for graduate students, and the number of awards given to geoscience graduates has been increasing since 2008 to a high of 148 awards in

2015 totaling $6.5 million in award money (Figure 3.55). However the number of geoscience awards dropped in 2016 to 109 totaling $5 million. The rapid increase in the total number of graduate fellowships was initially due to the ARRA stimulus funding in 2009, but the continuing high number of awards since then is due to a focus by NSF to do more to increase the quality of the future academic workforce.

The fields of study for the geoscience graduate fellowship awards indicate the popular areas of study for geoscience graduate students. Since 2012 there was a rapid increase in the number of awards in the geochemistry, paleoclimate, marine biology, and biogeochemistry fields (Figure 3.59).

Figure 3.54: Types of Financial Aid Used by Geoscience Graduates While Working Towards Their Degree, 2015

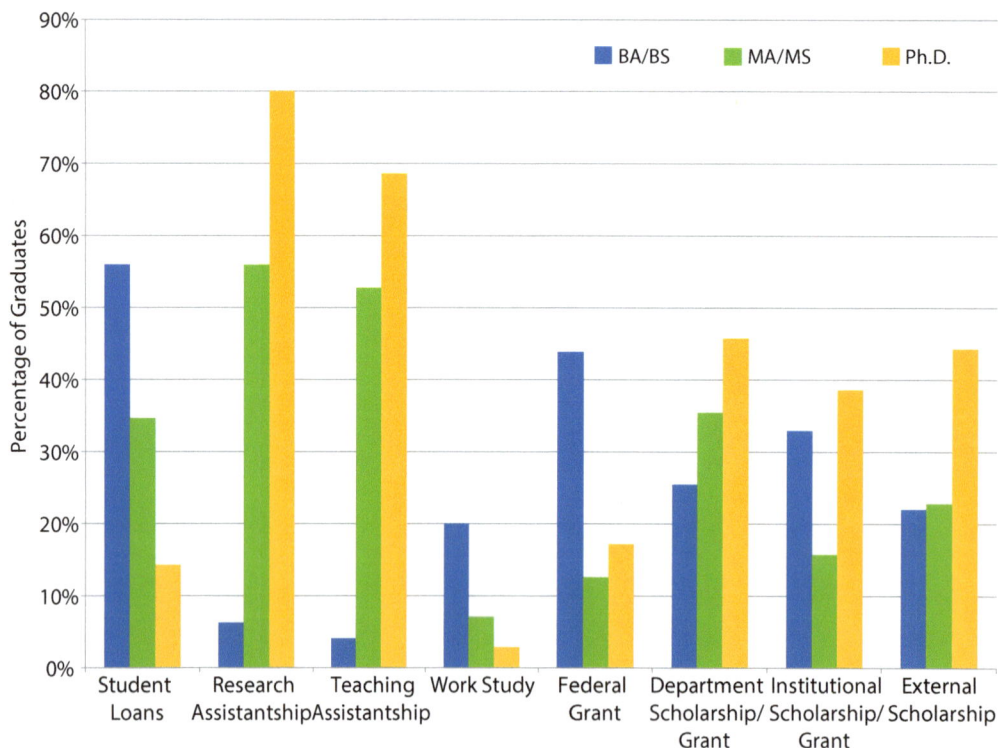

AGI Geoscience Workforce Program; Data derived from AGI's Geoscience Student Exit Survey 2015

Figure 3.55: Number of NSF Graduate Fellowships Awarded, 2007-2016

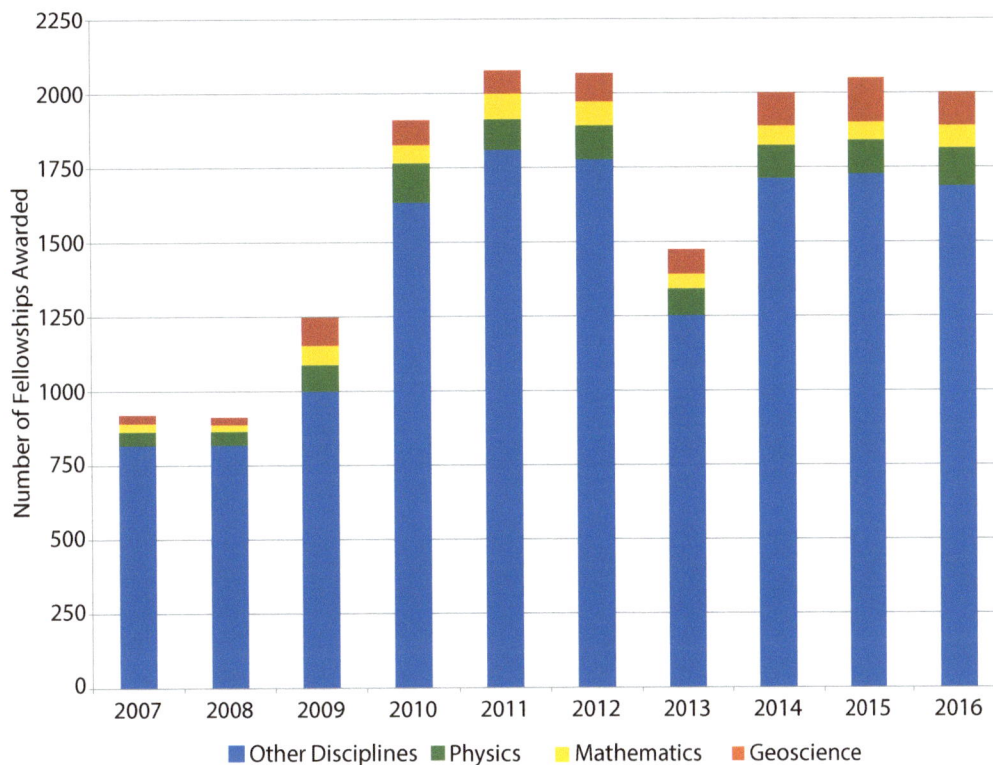

AGI Geoscience Workforce Program; Data derived from NSF Graduate Fellowship Program reports posted on data.gov

Figure 3.56: Total Funding of Geoscience NSF Graduate Fellowships

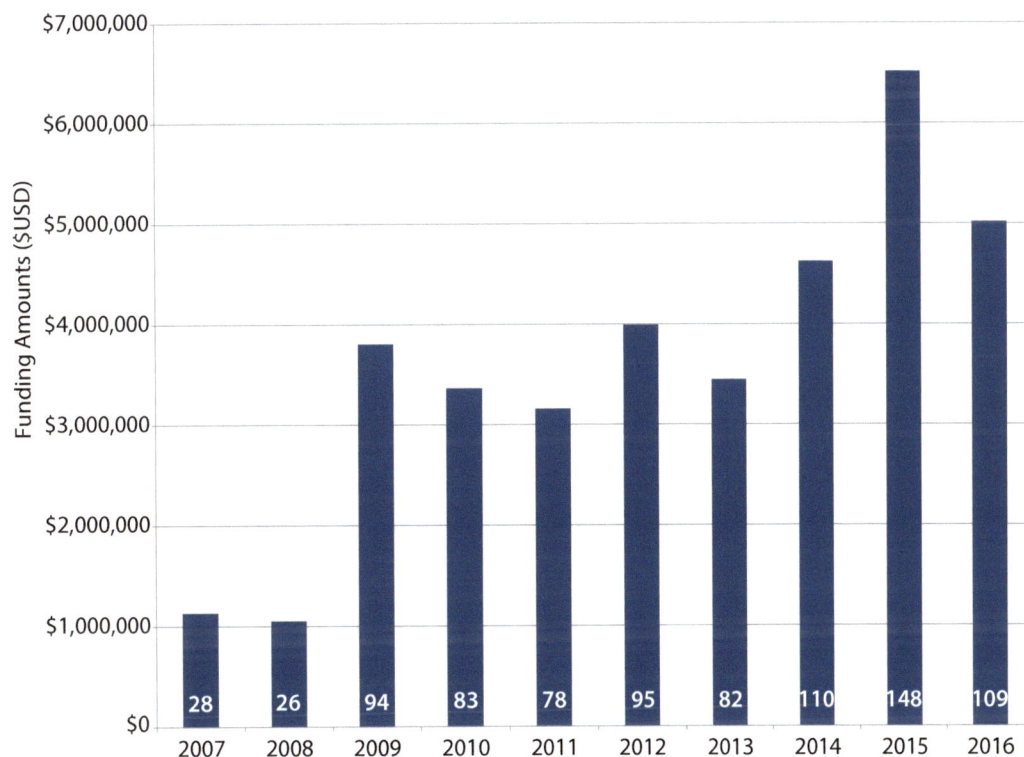

AGI Geoscience Workforce Program; Data derived from NSF Graduate Fellowship Program reports posted on data.gov

Figure 3.57: NSF Geoscience Graduate Fellowships by Field of Study, 2007-2016

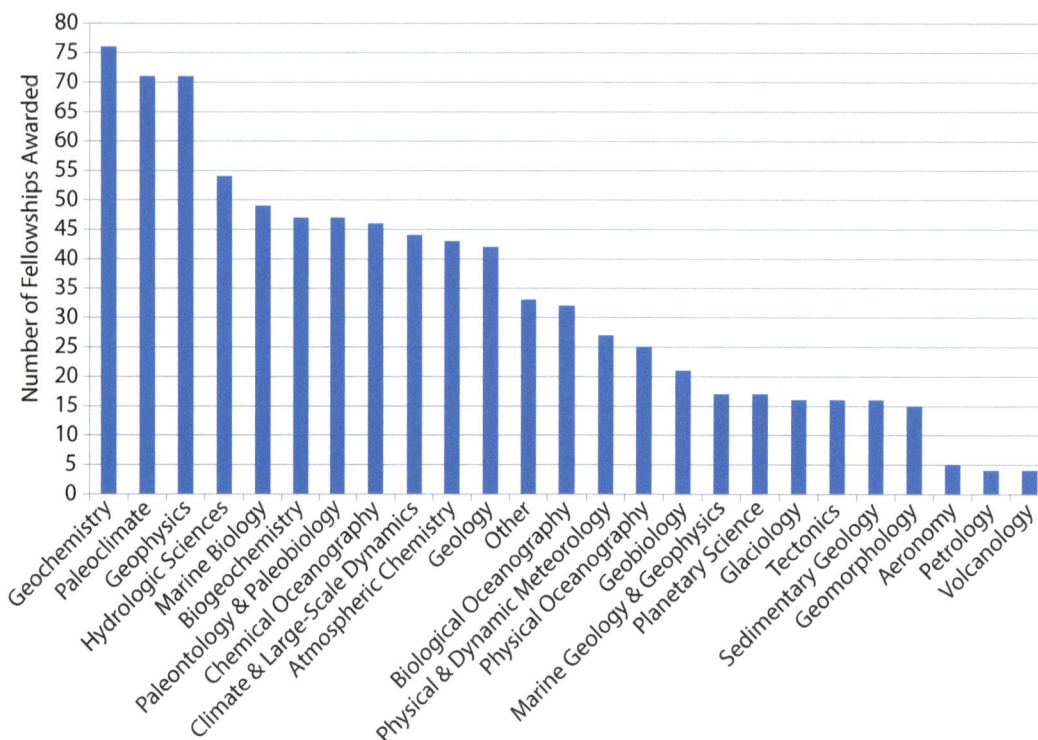

AGI Geoscience Workforce Program; Data derived from NSF Graduate Fellowship Program reports posted on data.gov

Table 3.14: Top 10 Baccalaureate Institutions Attended by NSF Geoscience Graduate Fellows (2007-2016)

Baccalaureate Institutions of Geoscience NSF Graduate Fellows	State	Number of Fellows (2007–2016)
Harvard University	MA	28
University of Washington	WA	25
Brown University	RI	23
Stanford University	CA	18
Massachusetts Institute of Technology	MA	17
University of California-Berkeley	CA	17
Columbia University	NY	15
University of Texas at Austin	TX	15
California Institute of Technology	CA	14
Cornell University	NY	14

AGI Geoscience Workforce Program; Data derived from NSF Graduate Fellowship Program reports posted on data.gov

Table 3.15: Top 10 Graduate Institutions Attended by NSF Geoscience Graduate Fellows (2007-2016)

Graduate Institutions of Geoscience NSF Graduate Fellows	State	Number of Fellows (2003-2012)
Massachusetts Institute of Technology	MA	41
University of Washington	WA	38
Harvard University	MA	27
University of California–San Diego	CA	26
Columbia University	NY	23
Stanford University	CA	22
University of California–Berkeley	CA	22
University of Colorado at Boulder	CO	21
California Institute of Technology	CA	20
University of Arizona	AZ	16

AGI Geoscience Workforce Program; Data derived from NSF Graduate Fellowship Program reports posted on data.gov

Chapter 4: Trends in Geoscience Employment — Examining Student Transitions and Workforce Dynamics

When discussing current and projected employment needs for the geosciences, the two major critical issues that face the geoscience workforce are the rate at which new talent transitions into geoscience professions and the rapid loss of experienced talent as they retire. Over the last several years, the improved employment prospects for geoscientists has spurred increase enrollments and awarded degrees within geoscience programs.

According to the Bureau of Labor Statistics (BLS), there were a total of 324,411 geoscience jobs in 2014, and this number is expected to increase by 10% by 2024 to a total of 355,862 jobs. Approximately, 156,000 geoscientists are expected to retire by 2024, but over the next decade, approximately 58,000 students will be graduating with their bachelor's, master's, or doctoral degrees in the geosciences and entering the geoscience workforce. Therefore, assuming minimal non-retirement attrition from the geoscience workforce, there is expected to be a deficit of approximately 90,000 geoscientists by 2024—a decrease from the previously predicted 135,000 geoscientists deficit. The projections based on the 2014 employment numbers present a smaller percentage increase in geoscience jobs over the next decade compared to the projections presented in 2012. Some of this can be attributed to changes in the oil and gas industry. The projected increase in jobs for petroleum engineers for the 2014-2024 time period was less than projected increase presented for the 2012-2022 time period. This decrease in the deficit of geoscience jobs can also be attributed to continued increases in enrollments and degrees awarded at the bachelor's and master's degree levels.

Over the past couple of years, there has been a steady downturn in the oil and gas industry, which has led to an increase in layoffs and a shifting in needs assessed within the industry. However, through discussions with industry representatives and through data collected by AGI, it appears that most of the layoffs occurred for mid-career and late-career employees. The industry is still hiring recent graduates at similar rates as in previous years, particularly among master's graduates. The environmental industry appears to be hiring more bachelor's graduates in 2015 than any other industry. Growth in the environmental industry can also be seen in the data provided by the Bureau of Labor Statistics through increases in employment in the environmental scientist and environmental engineer occupations. With multiple years of AGI's Exit Survey data, the connections between the various degree fields within the geosciences to industries hiring these graduates are becoming clearer and highlighting the diversity of career options for geoscience graduates.

While some industries and companies are working to bridge the future gap in the geoscience workforce, the federal and mining workforce are predicted to have negative growth in employment over the next decade with a decrease of 6% of mining jobs and 8% in federal jobs according to the Bureau of Labor Statistics. However, at least for the mining industry, this prediction of job loss might change as the industry conditions improve. The National Mining Association reported an increase in young geoscientists entering the mining industry and in support activities for mining and oil and gas, and economic indices have shown growth in the yield and value of minerals from U.S. mines. For some minerals, the number of mines in the U.S. has also increased.

Future research at AGI will be focused on tracking the early career geoscientists as they work to establish a permanent career. Careers are typically established five to seven years after completing a terminal degree. During this five year period, one can typically experience changes in personal, professional, and economic issues that will impact their workforce trajectory and lead to changes in jobs and/or industries. AGI plans to measure these changes in order to look at workforce skills development, compare entry-level positions with permanent positions, identify the various industries and organizations that hire geoscience graduates, and measure the attrition rate of early-career geoscientists from the geoscience workforce.

Early Career Workforce

Among geoscience graduates in 2015, 10% of bachelor's graduates, 40% of master's graduates, and 59% of doctoral graduates have obtained a job within the geosciences at graduation, and 51% of bachelor's graduates, 40% of master's graduates, and 39% of doctoral graduates are still looking for a job in the geosciences (Figure 4.1). This was the first year since starting AGI's Geoscience Student Exit Survey that an industry other than the oil and gas industry hired the highest percentage of bachelor's graduates (Figure 4.2). Approximately 40% of bachelor's graduates found a job in the environmental services industry, which was a 19% increase from 2013. The percentage of doctoral graduates hired by the oil and gas industry also decreased from 22% in 2013 to 15% in 2015. There is a decrease in master's graduates hired by the oil and gas industry between 2013 and 2015, but the oil and gas industry remains as the major hiring industry for this degree level. As the survey response rate increases and hiring rates change, different industries have appeared as options for students' right out of college, such as the non-profit industry, finance, and information services.

The starting salaries for graduates with geoscience degrees varies widely from less than $30,000 to more than $120,000; however, clear salary ranges can be seen for each degree level, with master's graduates tending to have the highest starting salaries (Figure 4.4). Geosciences graduates largely used their personal contacts, internet job searches, and faculty referrals to find these jobs, and 48% of master's graduates that found a job utilized campus recruiting events and job fairs very successfully (Figure 4.7).

The circular figure displays the connection between the degree fields of recent geoscience graduates from 2013-2015 (in color) to the industries where these geoscientists found their first job after graduation (in gray) (Figure 4.8). The size of the bars along the out edge of the circle represents the number of recent graduates that pursued a particular degree field and entered a particular industry. Each colored, inner ribbon connects a particular degree field with a job in a particular industry. This visualization shows the variety of industries available to graduates with a geoscience degree, as well as the complexity of the workforce and knowledge needed in the distinct industries.

Figure 4.1: Geoscience Graduates Seeking or Have Accepted a Position within the Geosciences, 2015

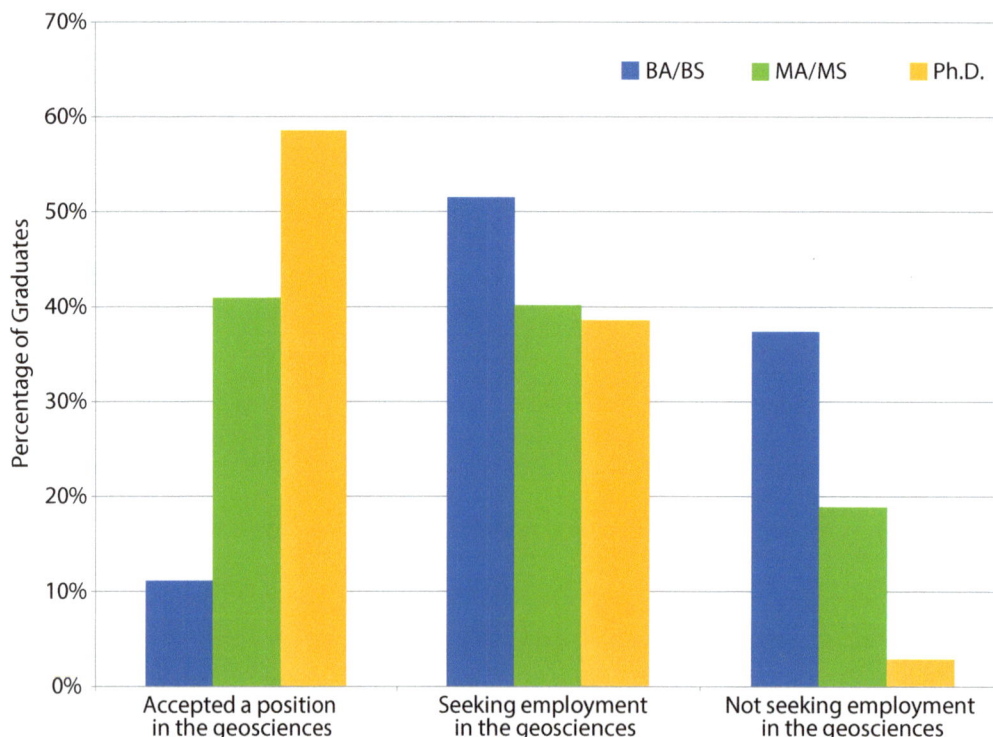

AGI Geoscience Workforce Program; Data derived from AGI's Geoscience Student Exit Survey 2015

Figure 4.2: Industries Hiring Geoscience Graduates, 2015

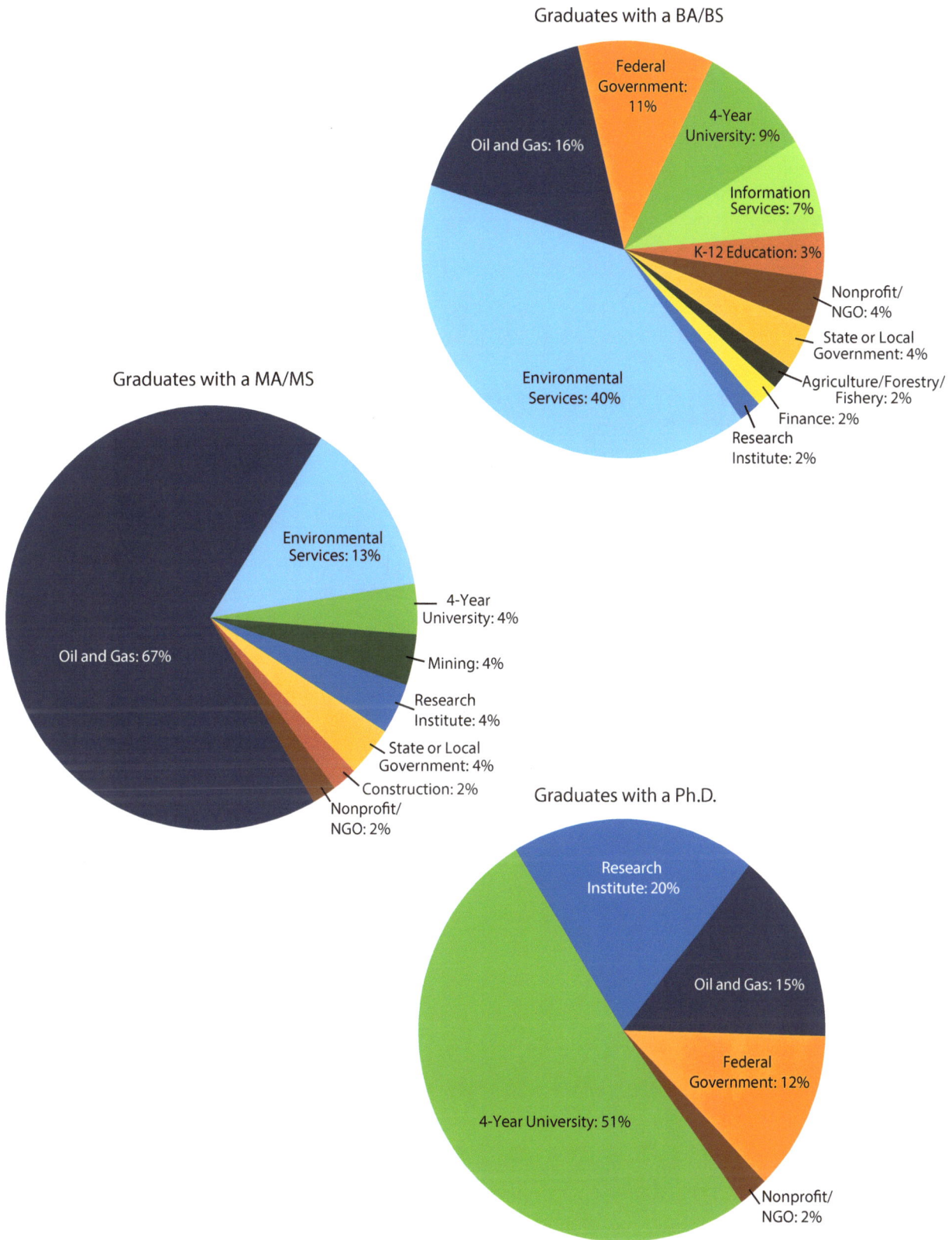

Graduates with a BA/BS

Graduates with a MA/MS

Graduates with a Ph.D.

AGI Geoscience Workforce Program; Data derived from AGI's Geoscience Student Exit Survey

Figure 4.3: Industries of Interest for Graduating Students Seeking a Job within the Geosciences, 2015

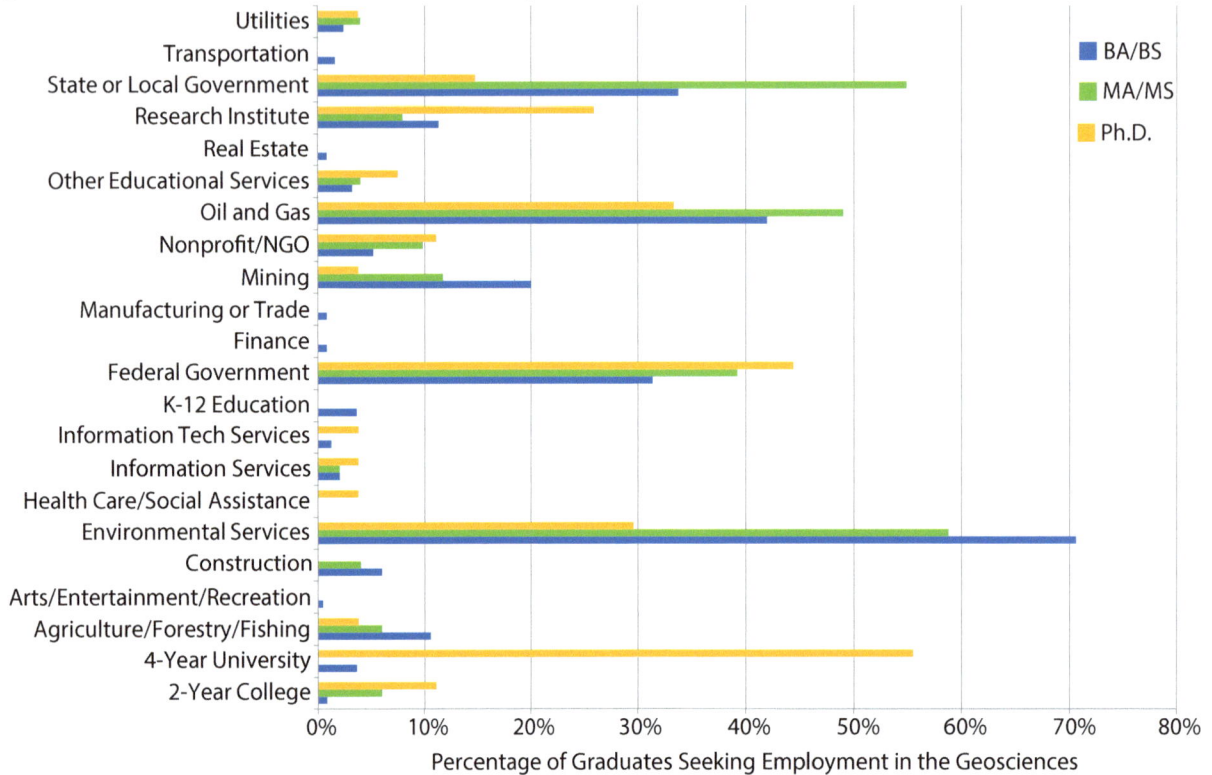

AGI Geoscience Workforce Program; Data derived from AGI's Geoscience Student Exit Survey 2015

Figure 4.4: Starting Salaries for Employed Geoscience Graduates, 2015

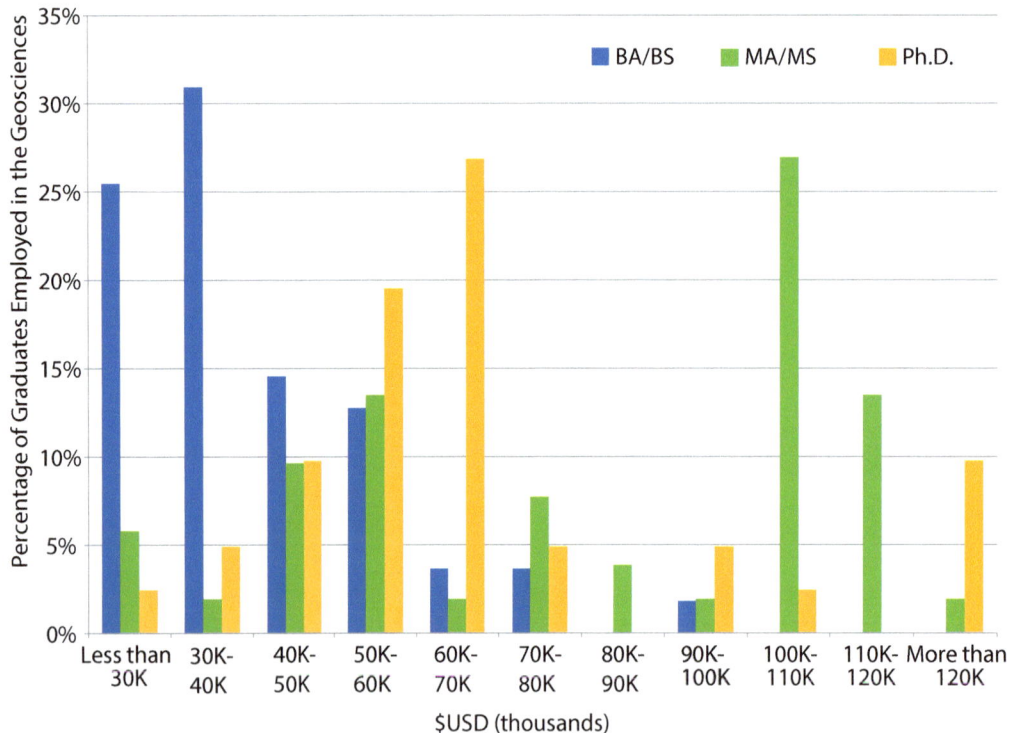

AGI Geoscience Workforce Program; Data derived from AGI's Geoscience Student Exit Survey 2015

Figure 4.5: Additional Compensation Granted to Geoscience Graduates with a Geoscience Job, 2015

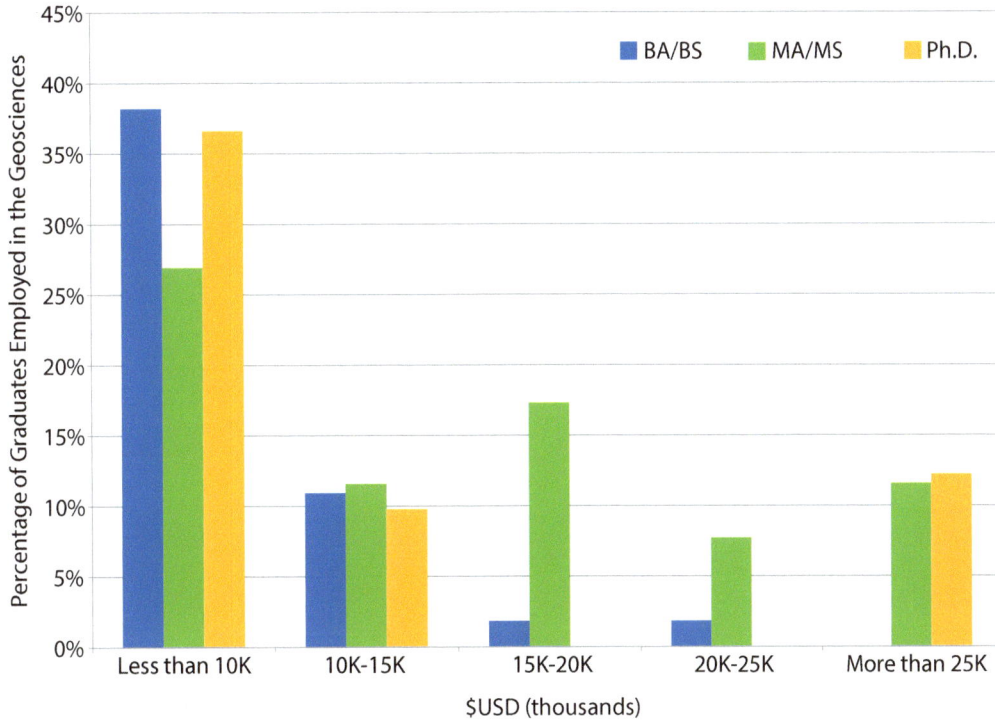

AGI Geoscience Workforce Program; Data derived from AGI's Geoscience Student Exit Survey 2015

Figure 4.6: Other Job Opportunities Granted to Employed Geoscience Graduates, 2015

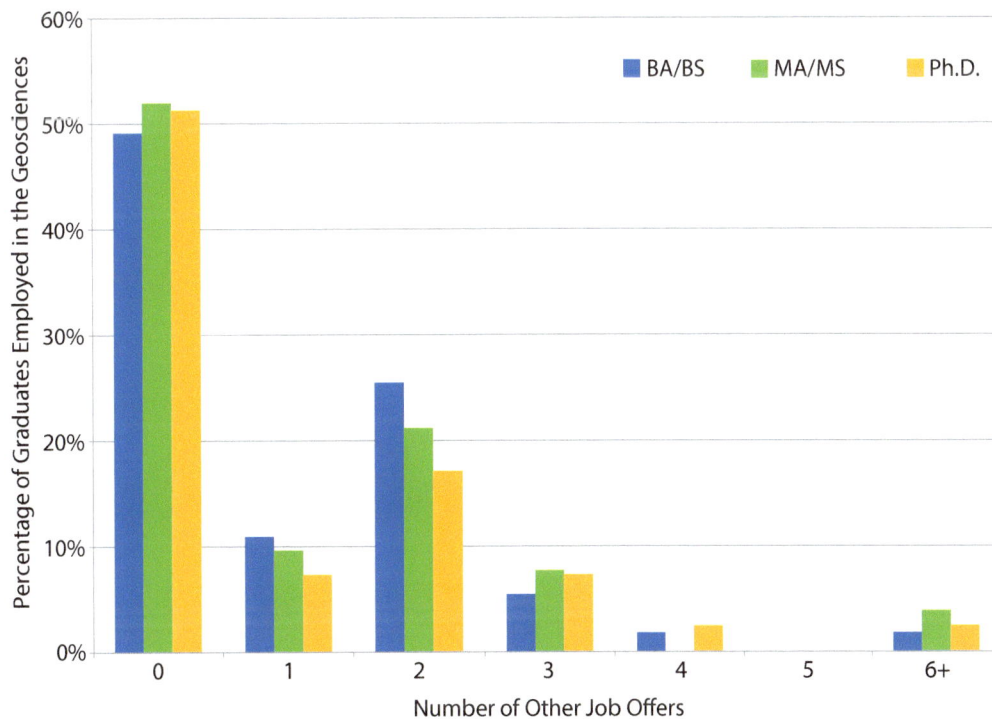

AGI Geoscience Workforce Program; Data derived from AGI's Geoscience Student Exit Survey 2015

Figure 4.7: Useful Resources Used by Geoscience Students to Find a Job, 2015

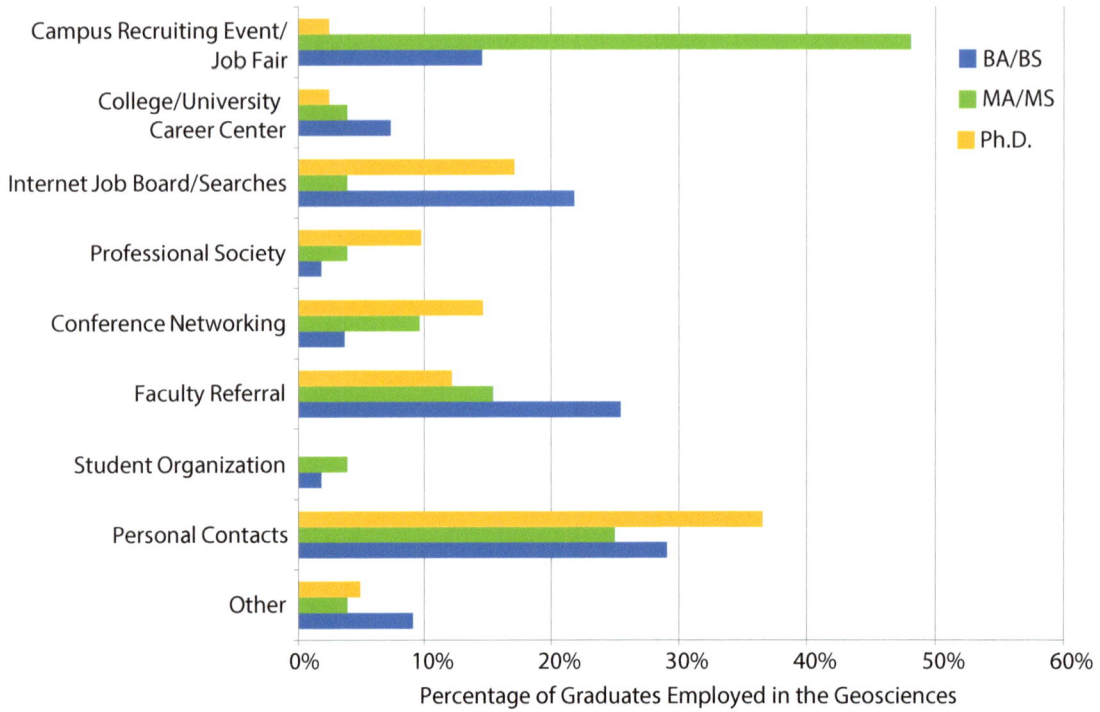

AGI Geoscience Workforce Program; Data derived from AGI's Geoscience Student Exit Survey 2015

Figure 4.8: Industries of Geoscience Graduates' First Jobs by Degree Field (2013-2015)

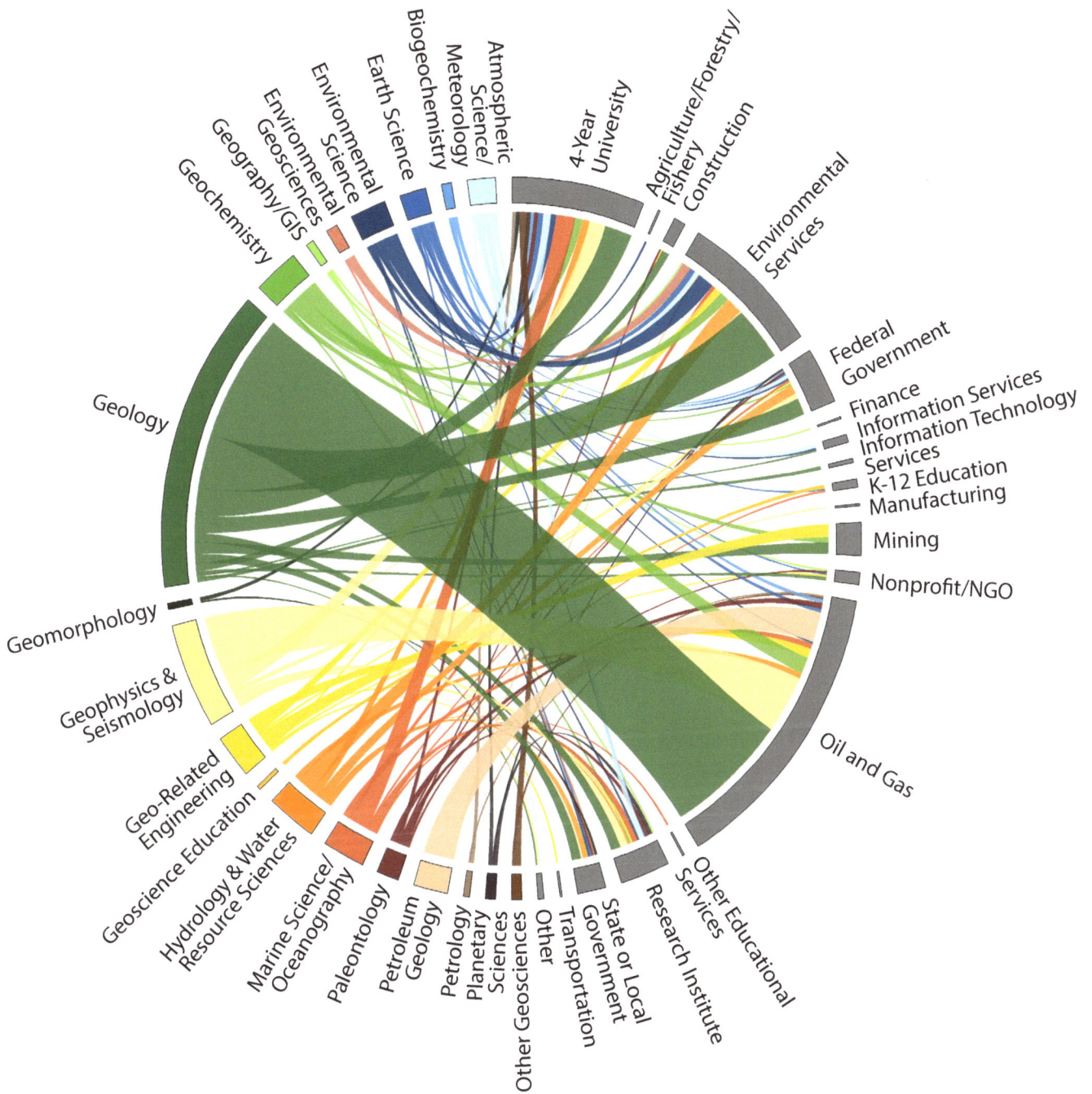

AGI Geoscience Workforce Program; Data derived from AGI's Geoscience Student Exit Survey 2015

Salary Trends for Geoscience Occupations

Median annual salaries present a more realistic salary for a particular occupation compared to mean annual salaries. Geoscience salaries have increased by 3% since 2012, which was similar to the growth seen in other science occupations (3%) and slightly lower than all U.S. occupations (4%) (Figure 4.9). In 2015, the geoscience occupations with the highest median salaries were for engineering managers ($132,800), petroleum engineers ($129,990), natural science managers ($120,160), and mining and geological engineers ($94,040). Figure 4.10 shows the median salaries of geoscience occupations and larger groups of occupations for 2015 to compare the geoscience occupational salaries within the major groups with the median of all the occupations within the major groups. All the geoscience occupations have a median salary above the overall median salary in the United States for all occupations. The geoscience workforce continues to be a lucrative career field in the United States.

Figure 4.9: Median Annuals Salaries of Geoscience Occupations (2005-2015)

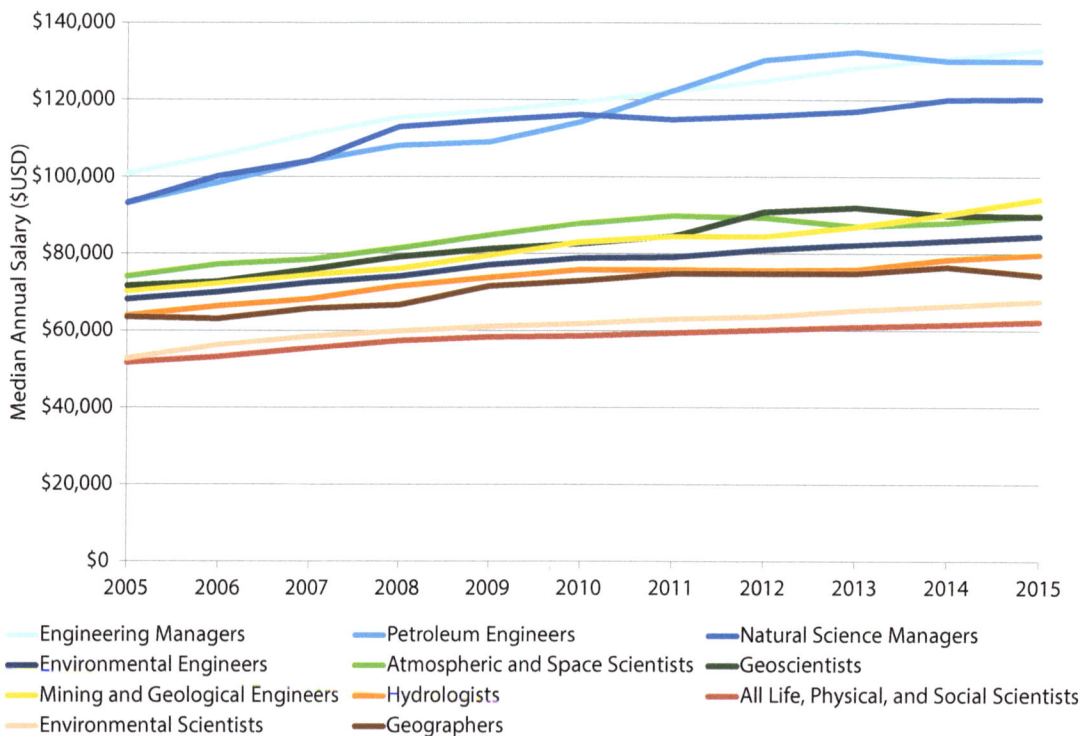

AGI Geoscience Workforce Program; Data derived from the U.S. Bureau of Labor Statistics, National Occupational Employment and Wage Estimates

Figure 4.10: Median Annual Salaries of Geoscience Occupations, 2015

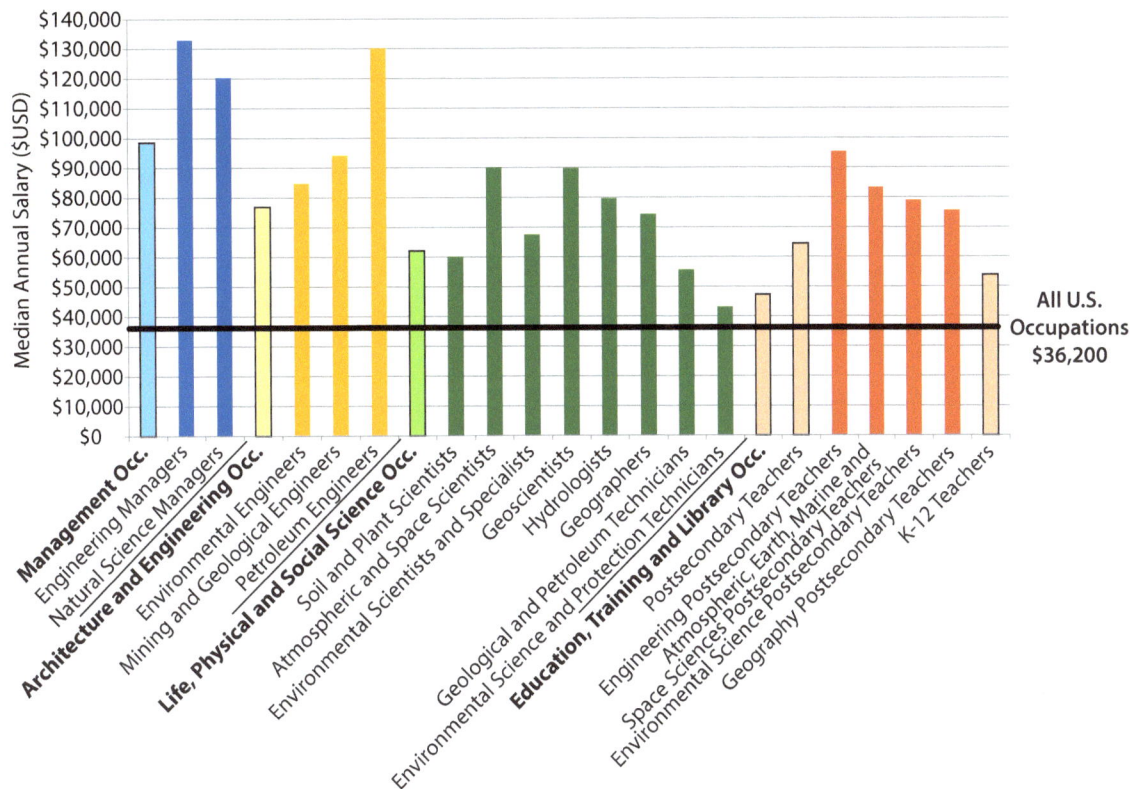

AGI Geoscience Workforce Program; Data derived from the U.S. Bureau of Labor Statistics, National Occupational Employment and Wage Estimates

Demographics of the Geoscience Profession

While it looks like the total number of geoscientists employed in the United States has increased from 2012-2014, in reality, the 2014 number of geoscientists include the number of postsecondary teachers in atmospheric sciences, earth sciences, marine sciences, space sciences, environmental sciences, and geography fields as counted by the Bureau of Labor Statistics (Figure 4.9). When reviewing the list of occupations by the Bureau of Labor Statistics, it became clear that these occupations should be included in the counts of total geoscientists. If the counts of the postsecondary teachers are removed from the 2014 number of geoscientists, the change in the number of geoscientists in the workforce from 2012 to 2014 was 0.7%. With the postsecondary teachers included, there were 324,411 geoscientists working in the United States according to the Bureau of Labor Statistics. A large majority of the geoscientists working in the U.S. have occupations within the professional, scientific, and technical services industry. This industry includes occupations related to research facilities, testing laboratories, and architectural engineering (Figure 4.12). The majority of geoscientists tend to fall into the environmental scientist and environmental engineer occupation categories (Figure 4.13).

The percentage of female geoscientists has hovered around 25%, and the highest percentage of female geoscientists are environmental engineers and oceanographers (Figures 4.14-4.16).

There is a discrepancy between the data on underrepresented minorities in the geoscience workforce from the Bureau of Labor Statistics and the National Science Foundation. According to the BLS, the percentage of underrepresented minority geoscientists has hovered between 8-9% from 2010-2015 (Figure 4.17). However, the NSF reported that 13% of geoscientists were from underrepresented minority groups in 2013 (Figure 4.18). This variance is likely a result from the changes in race/ethnicity classification that occurred in 2010.

Figure 4.11: Total Number of Employed Geoscientists in the United States

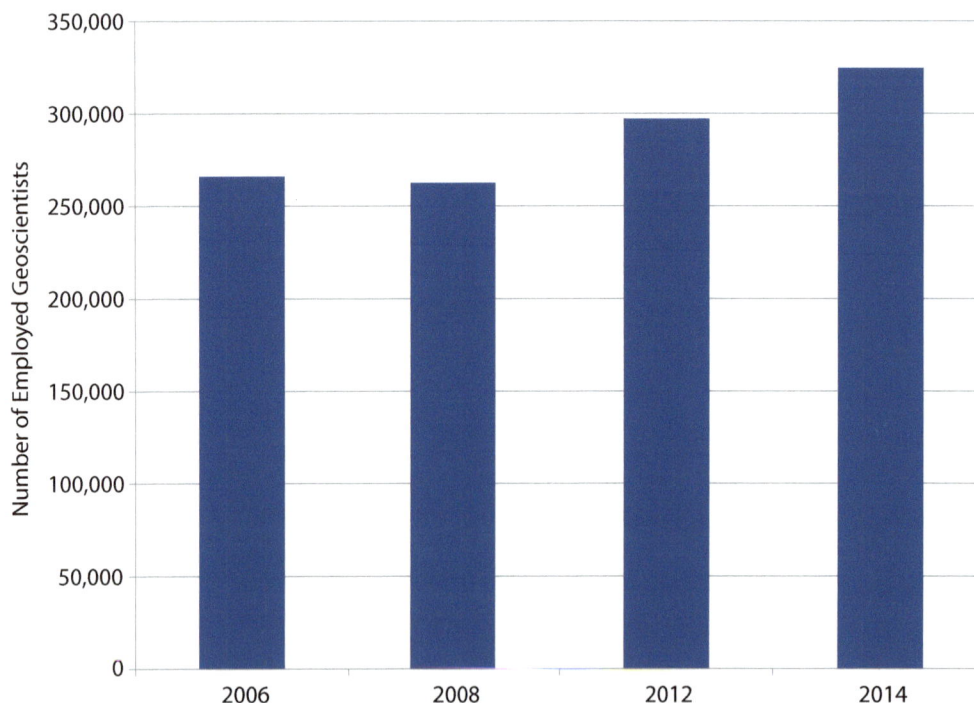

AGI Geoscience Workforce Program; Data derived from the U.S. Bureau of Labor Statistics, Employment Projections

Figure 4.12: Number of Geoscience Jobs by Industry Sector in 2014

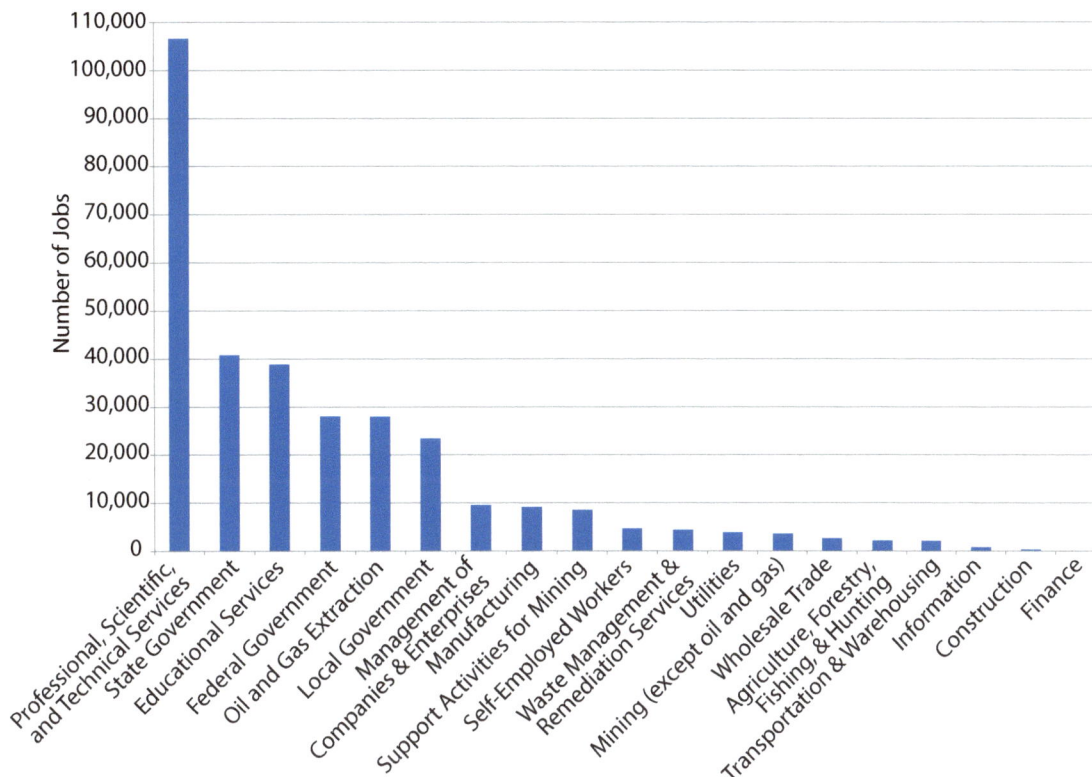

AGI Goescience Workforce Program, Data derived from the U.S. Bureau of Labor Statistics, Employment Projections

Figure 4.13: Current Employment for Detailed Geoscience Occupations (2008-2014)

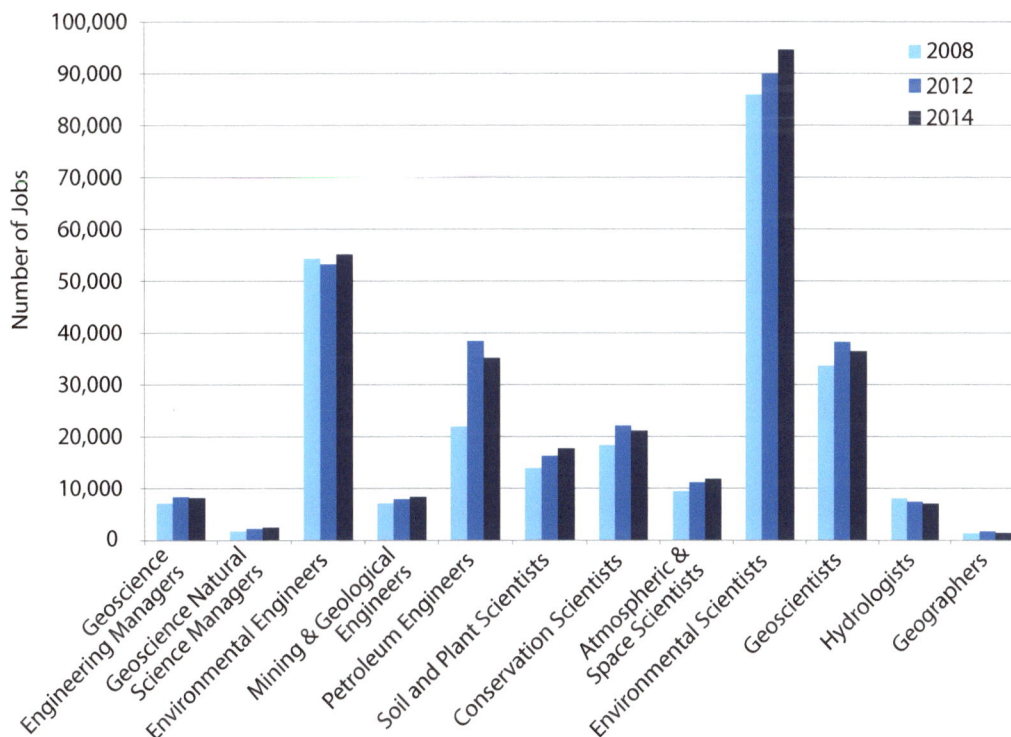

AGI Geoscience Workforce Program; Data derived from the U.S. Bureau of Labor Statistics, Employment Projections

Figure 4.14: Percentage of Women in Environmental Science and Geoscience Occupations

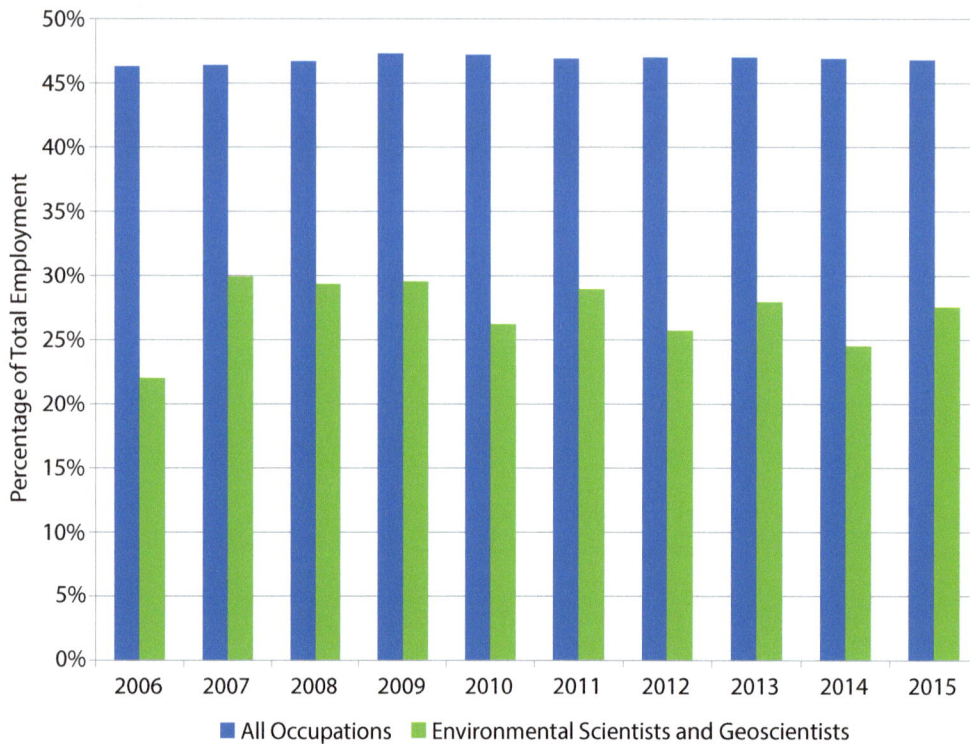

AGI Geoscience Workforce Program; Data derived from the U.S. Bureau of Labor Statistics, Current Population Survey

Figure 4.15: Percentage of Women in Geoscience and Other Science and Engineering Occupations, 2013

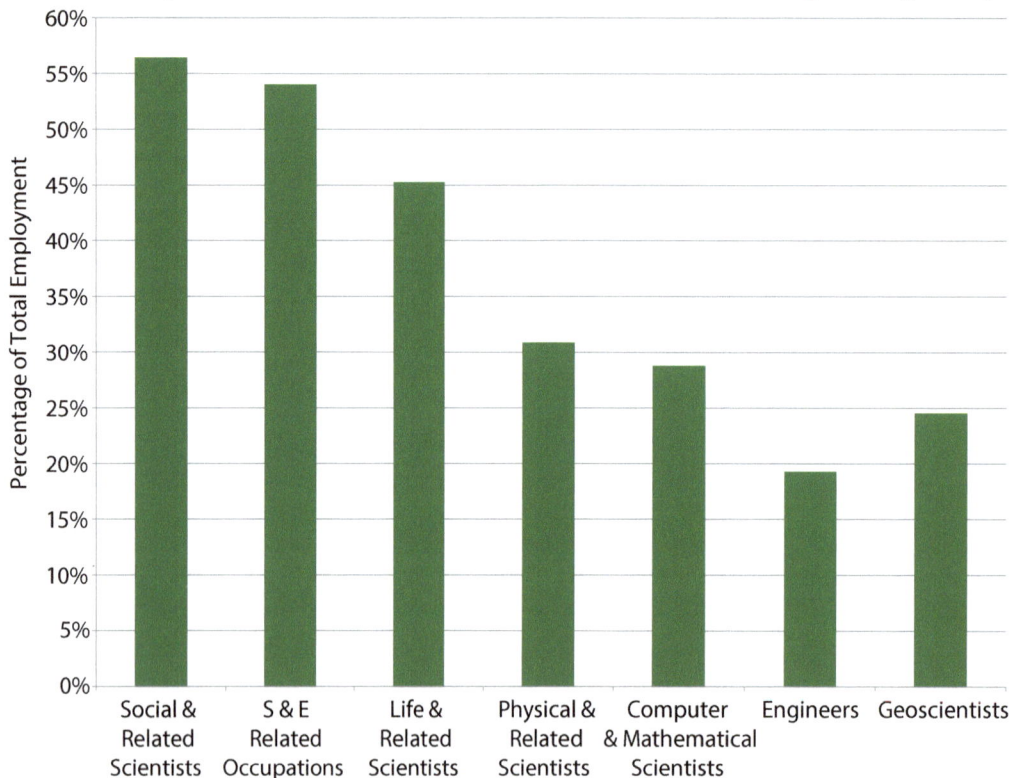

AGI Geoscience Workforce Program; Data derived from NSF's SESTAT Restricted-Use Data files, 2013

Figure 4.16: Percentage of Women in Detailed Geoscience Occupations, 2013

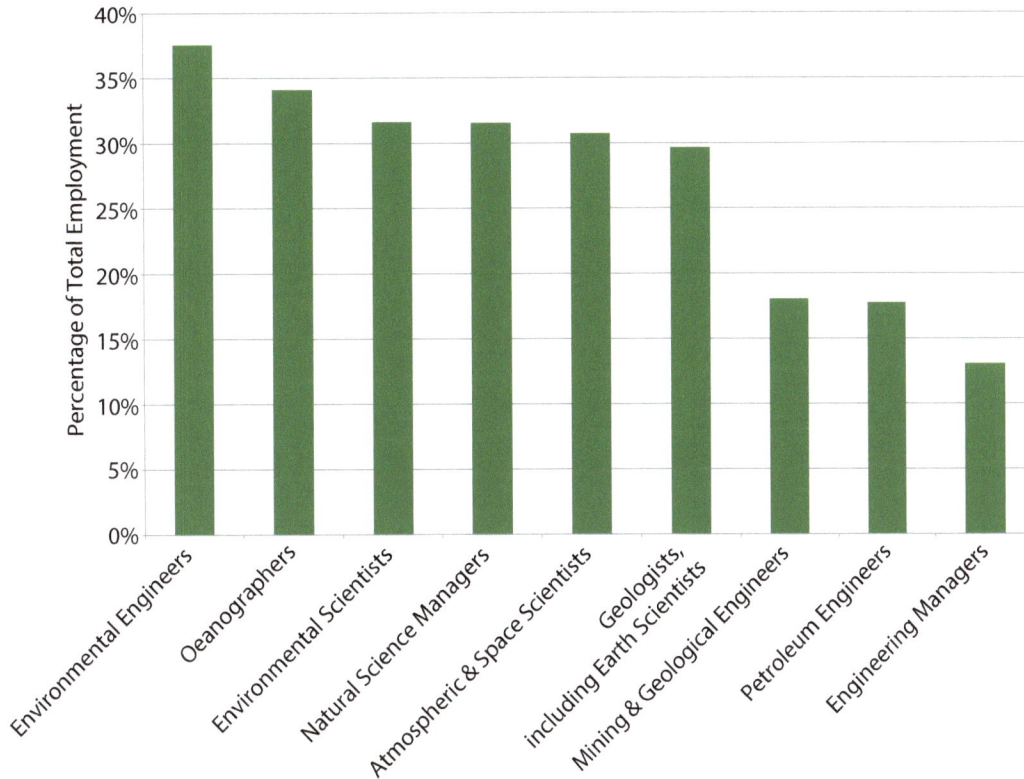

AGI Geoscience Workforce Program; Data derived from NSF's Restricted-Use Data files, 2013

Figure 4.17: Percentage of Underrepresented Minorities in Environmental Science and Geoscience Occupations

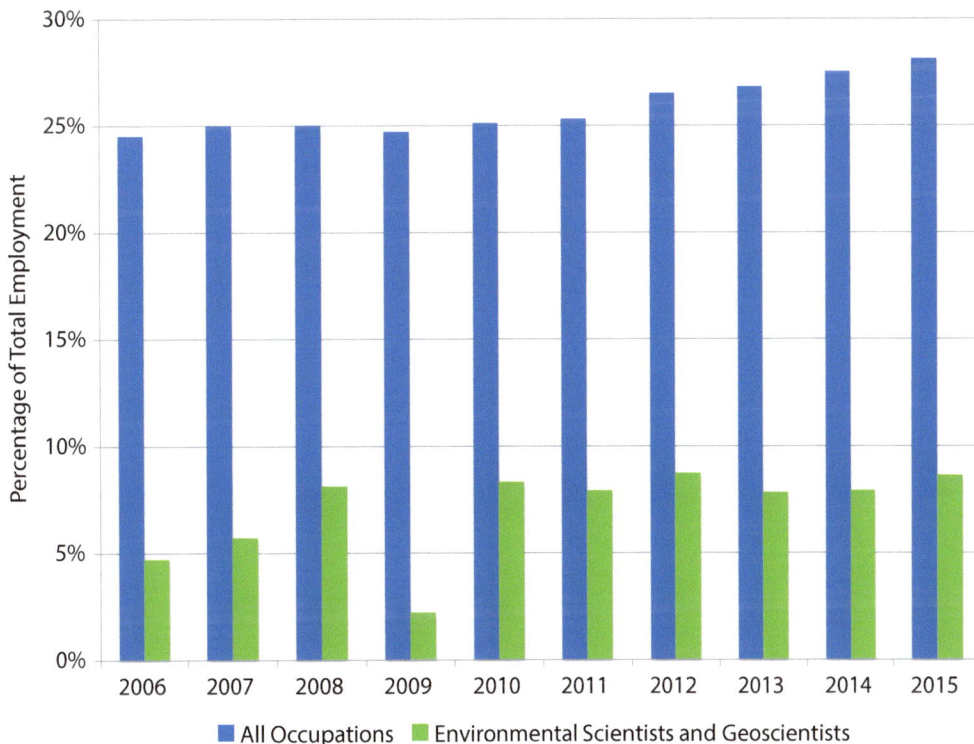

AGI Geoscience Workforce Program; Data derived from the U.S. Bureau of Labor Statistics, Current Population Survey

Figure 4.18: Percentage of Underrepresented Minorities in Geoscience and Other Science and Engineering Occupations, 2013

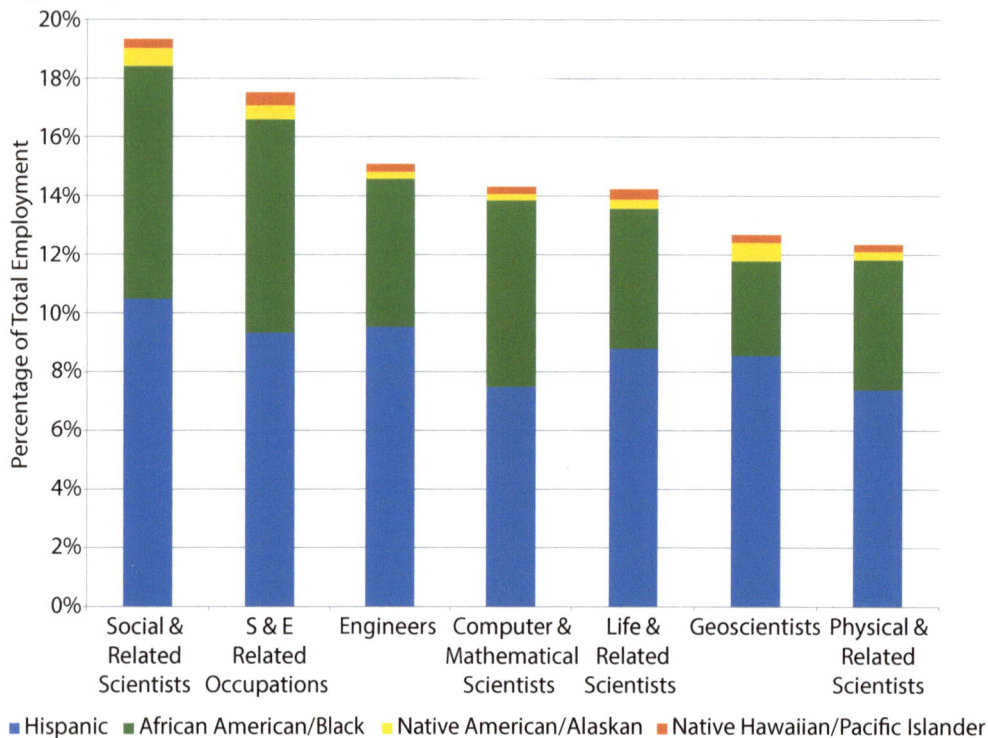

AGI Geoscience Woirkforce Program; Data derived from NSF's SESTAT Restricted-Use Data files, 2013

Figure 4.19: Percentage of Underrepresented Minorities in Detailed Geoscience Occupations, 2013

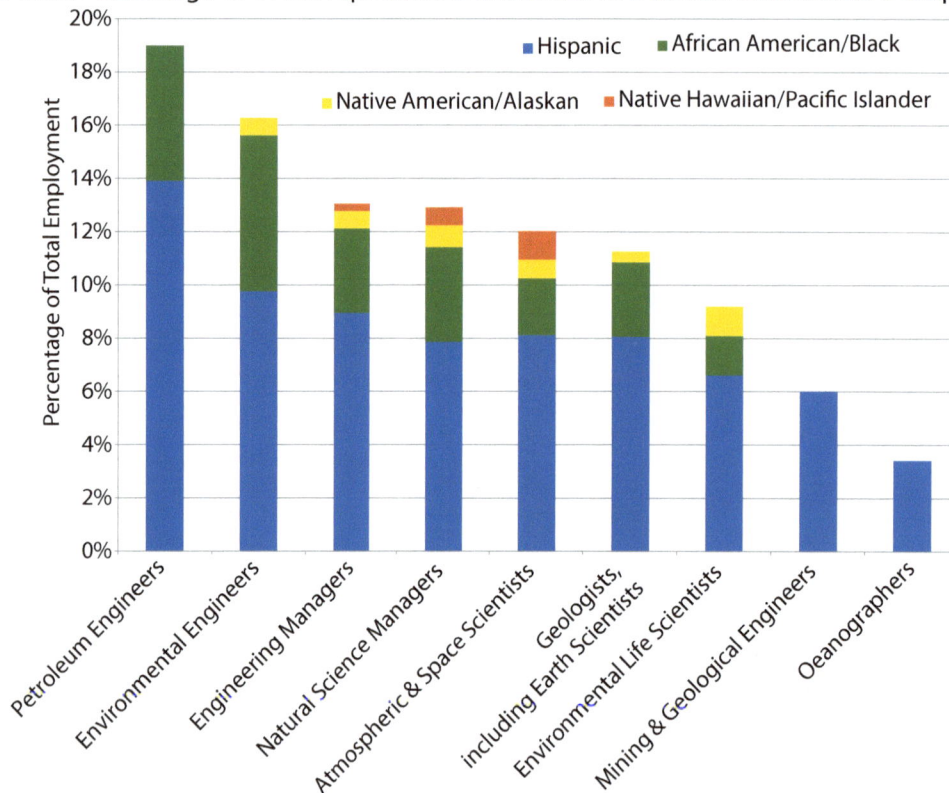

AGI Geoscience Workforce Program; Data derived from NSF's SESTAT Restricted-Use Data files, 2013

Workforce Age Demographics

The age distribution of the members of professional societies, such as the American Association of Petroleum Geologists, the Society of Economic Geologists, the Society of Exploration Geophysicists, and the National Groundwater Association, provide a good representation of the general age distribution of the traditional geoscience workforce. Figures 4.20 and 4.21 present the same data, but Figure 4.20 displays the society membership percentage for their members that are 30 and under. Many geoscience professional societies are actively trying to engage with students and increase their student membership, and they appear to be successful in this effort. When the student membership data is ignored, the majority of the traditional geoscience workforce is in their late 50's, except the majority of the hydrologists tend to be in their late 40's (Figure 4.17). The high percentages seen for the age group 31-40 may still have some artifact of student memberships, but it is also high because it covers 10 years, whereas most of the age groups cover geoscientists in 5 year groups.

The age distributions of geoscientists in the federal government are trending towards retirement age without clear signs of future replacements for the aging workforce (Figures 4.22-4.30). In fact, the overall employment of geoscientists in the federal government has decreased by 9% since 2009.

The model predicting the supply and demand of employees in the petroleum industry was updated taking into account recent data AGI has collected showing the influx of new graduates into the petroleum industry in 2015. Because AGI believes that most departments are at capacity with student enrollments, a realistic one percent growth in geoscience graduate students entering the petroleum industry was used to represent the current and future workforce, compared to the demand that will develop as the current workforce reaches retirement. With these changes to the model, it appears the petroleum industry has been working to bridge the future gap in the workforce that may appear as the majority of the workforce reaches retirement age (Figure 4.31). It is important to note that this model assumes that the majority of new graduates that enter the petroleum industry remain within the industry for a full career, which may not be the case. This model also does not take into account the recent layoffs in the oil and gas industry. The total current workforce line may be a bit lower and growth of the industry may change the shape of the predicted demand.

According to the National Mining Association, the mining industry continues to have had an influx of young geoscientists for all mining (except oil and gas extraction) and for support activities for mining and oil and gas (Figure 4.32 and 4.33). This is in agreement with workforce discussions with the mining industry. This industry seems to be a growing area of the geoscience workforce recently.

Between 2009 and 2015, there has been minimal change in the percentage distribution of faculty by rank with changes only covering a percentage point or two (Figure 4.34). Between 2013 and 2015 there has been a slight shift in the age distribution of assistant and associate professors to slightly younger in 2015, but the other ranks appear to have similar age distributions (Figure 4.35). There appears to be small hiring gains with 28 more assistant professors in 2015 than in 2013. Progress of tenure-track faculty moves steadily through the ranks to reach full professor between the ages of 46-50 on average. However, full professors tend to work later in their career creating a crossover in the population of full professors and emeritus in their 70's.

Figure 4.20: Geoscience Age Distribution by Member Society

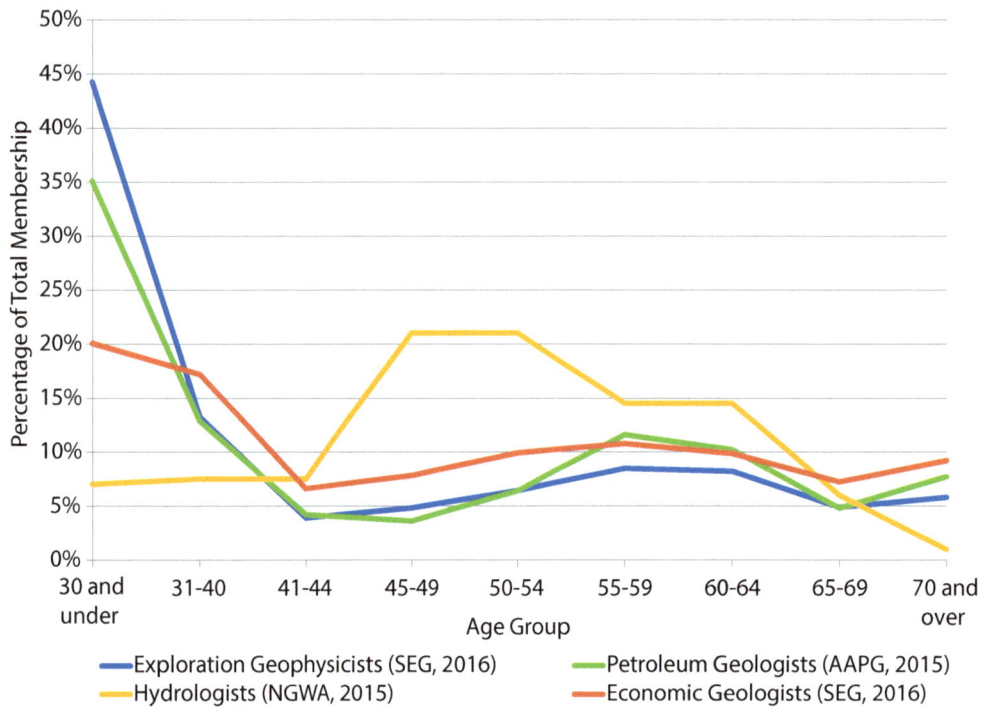

AGI Geoscience Workforce Program; Data provided by the Society of Exploration Geophysicists, American Association of Petroleum Geologists, Society of Economic Geologists, and the National Groundwater Association

Figure 4.21: Geoscience Age Distribution by Membership Society without Student Memberships

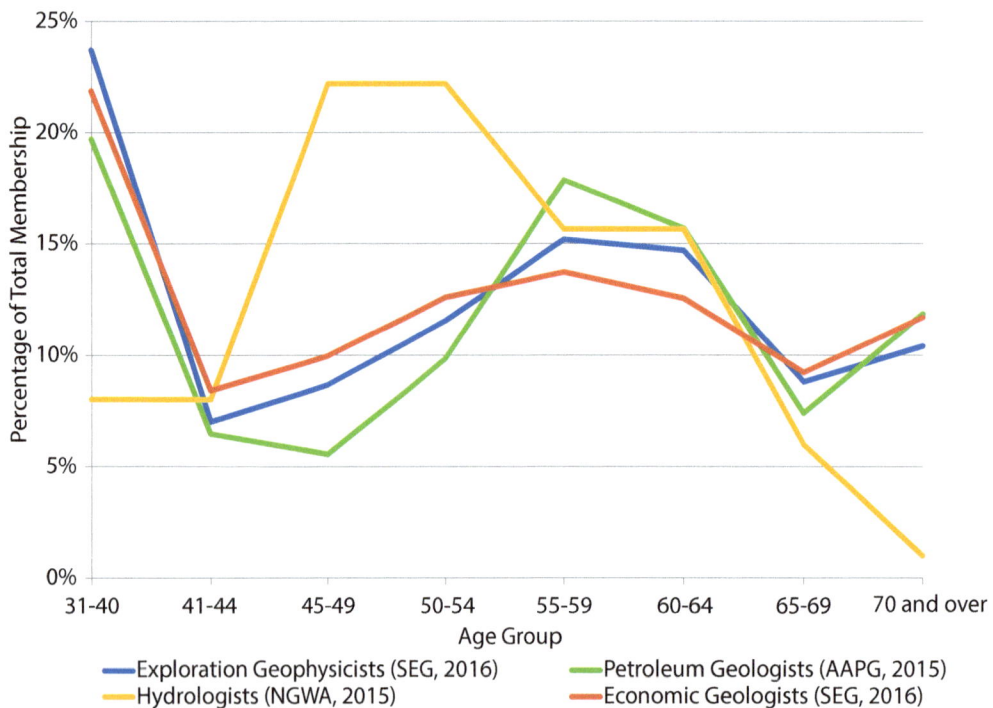

AGI Geoscience Workforce Program; Data provided by the Society of Exploration Geophysicists, American Association of Petroleum Geologists, Society of Economic Geologists, and the National Groundwater Association

Figure 4.22: Age Distribution of Geoscientists in the U.S. Government

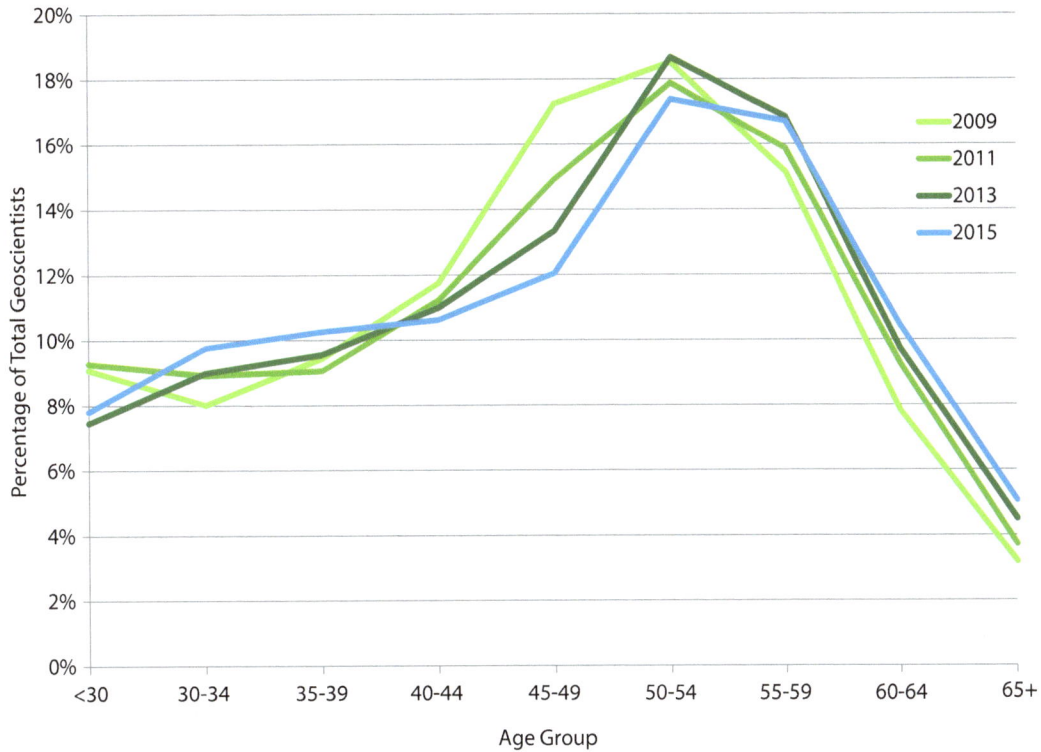

AGI Geoscience Workforce Program; Data derived from the Office of Personnel Management fedscope database

Figure 4.23: Age Distribution of Environmental Engineers in the U.S. Government

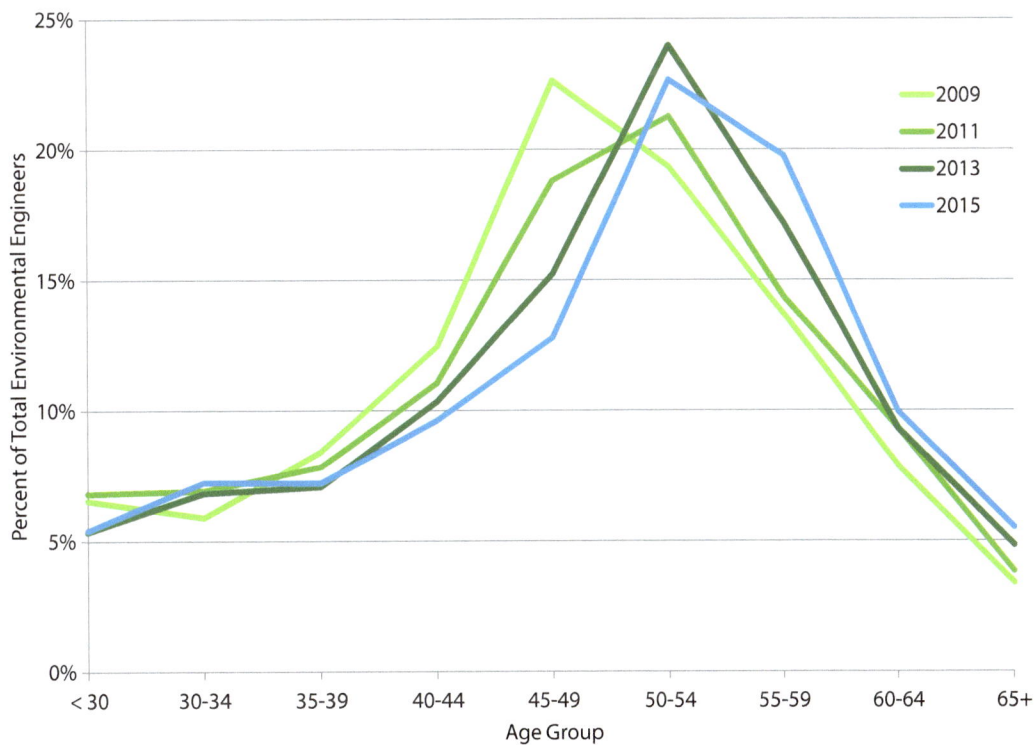

AGI Geoscience Workforce Program; Data derived from the Office of Personnel Management fedscope database

Figure 4.24: Age Distribution of Mining Engineers in the U.S. Government

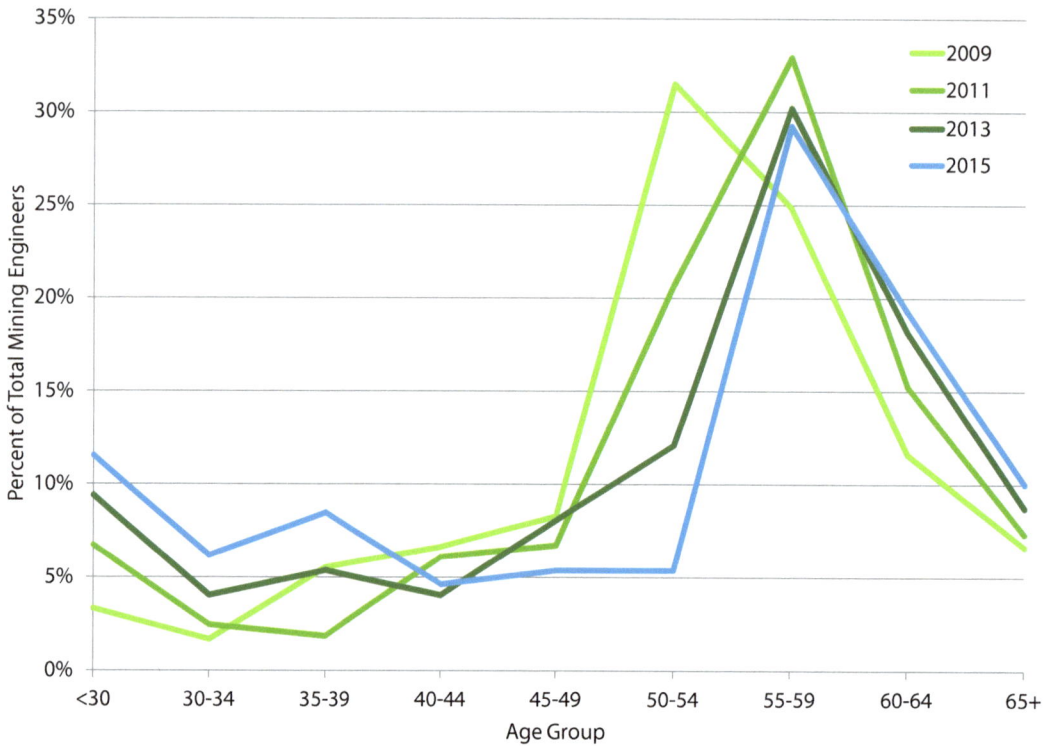

AGI Geoscience Workforce Program; Data derived from the Office of Personnel Management fedscope database

Figure 4.25: Age Distribution of Geophysicists in the U.S. Government

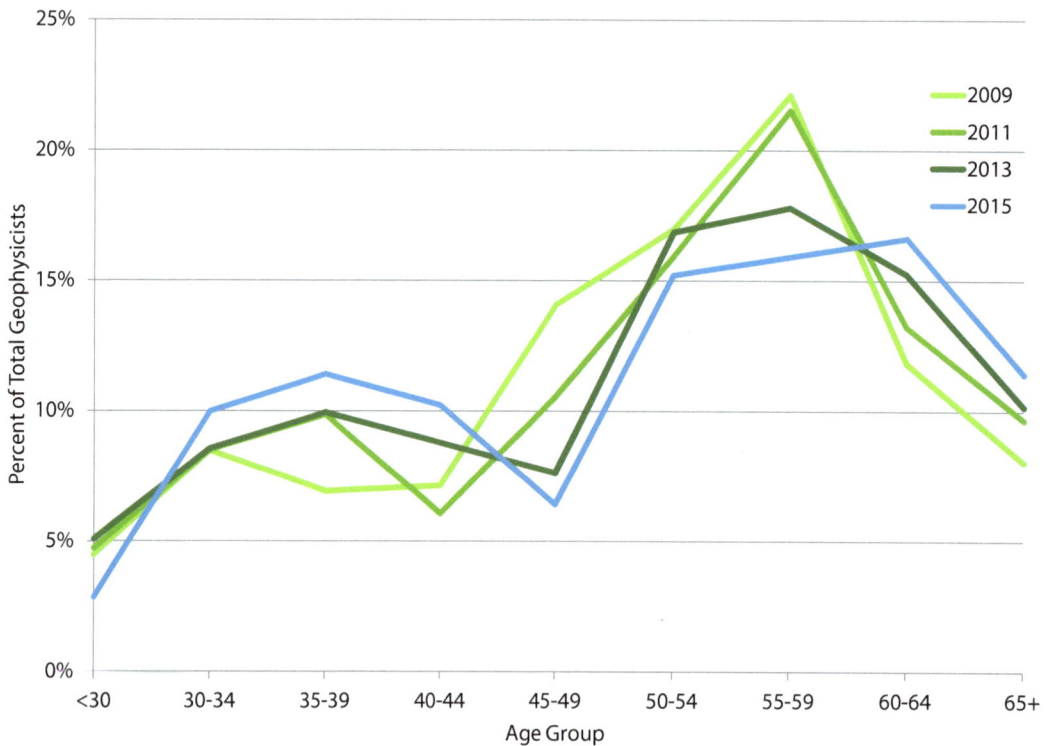

AGI Geoscience Workforce Program; Data derived from the Office of Personnel Management fedscope database

Figure 4.26: Age Distribution of Hydrologists in the U.S. Government

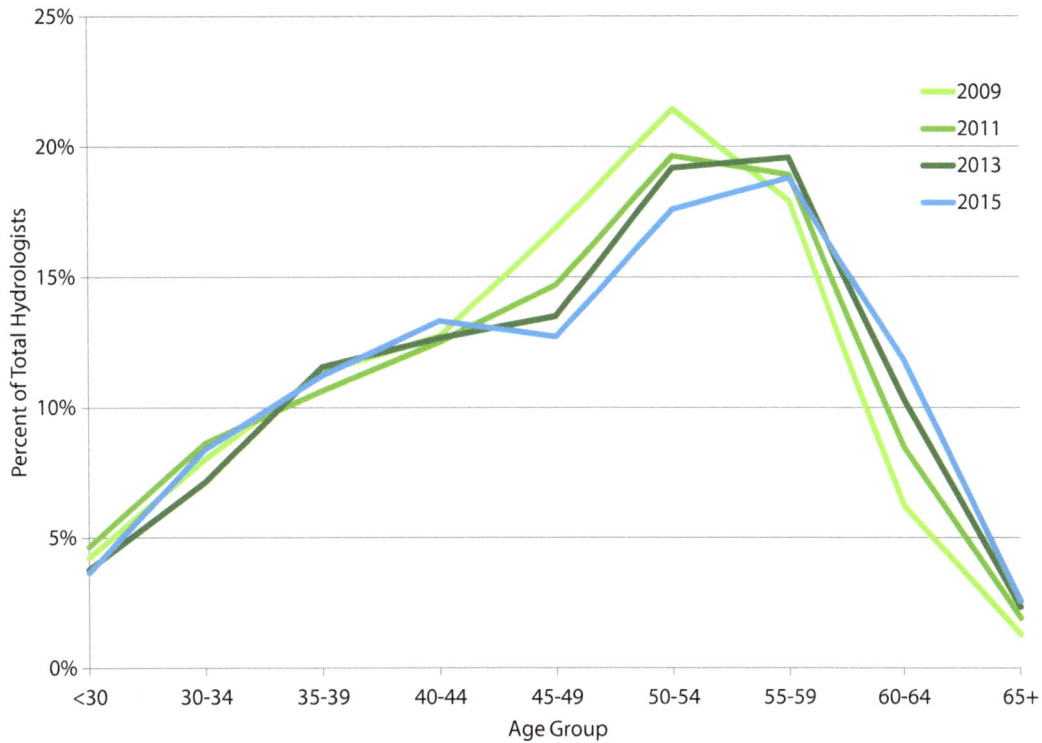

AGI Geoscience Workforce Program; Data derived from the Office of Personnel Management fedscope database

Figure 4.27: Age Distribution of Meteorologists in the U.S. Government

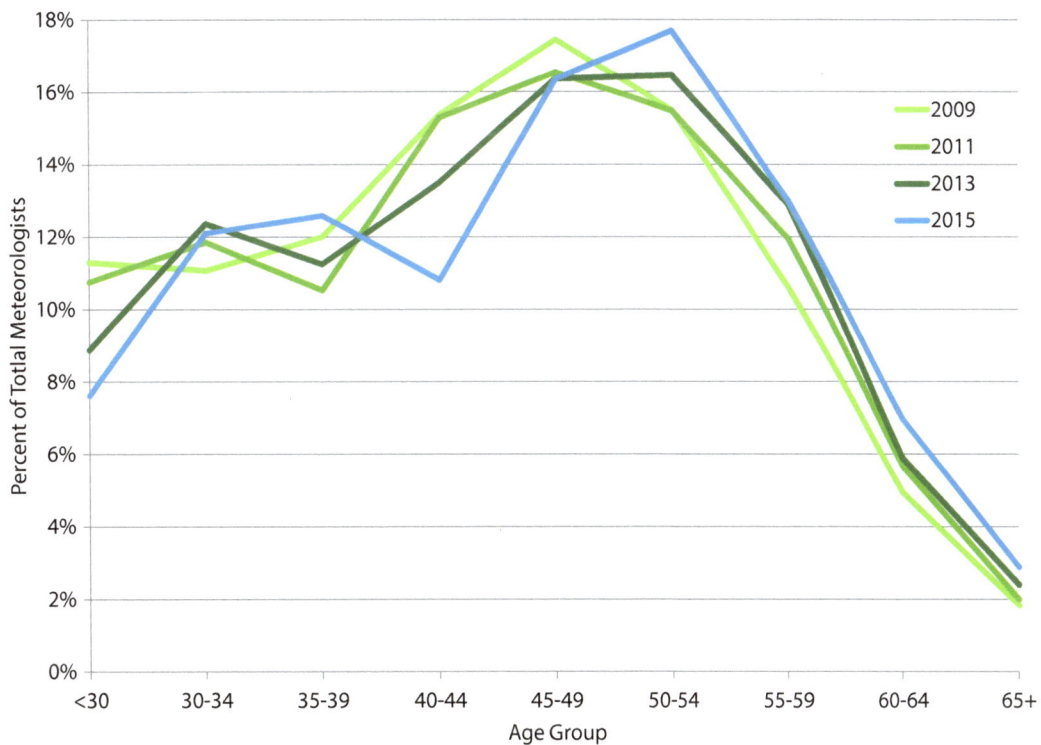

AGI Geoscience Workforce Program; Data derived from the Office of Personnel Management fedscope database

Figure 4.28: Age Distribution of Geologists in the U.S. Government

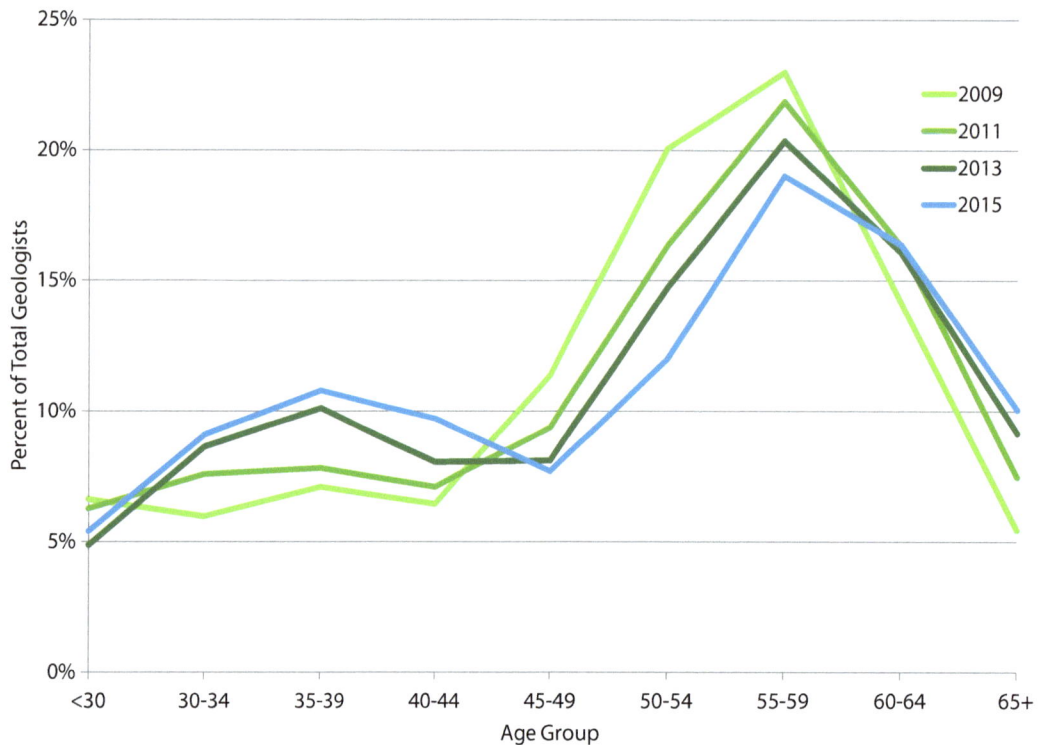

AGI Geoscience Workforce Program; Data derived from the Office of Personnel Management fedscope database

Figure 4.29: Age Distribution of Oceanographers in the U.S. Government

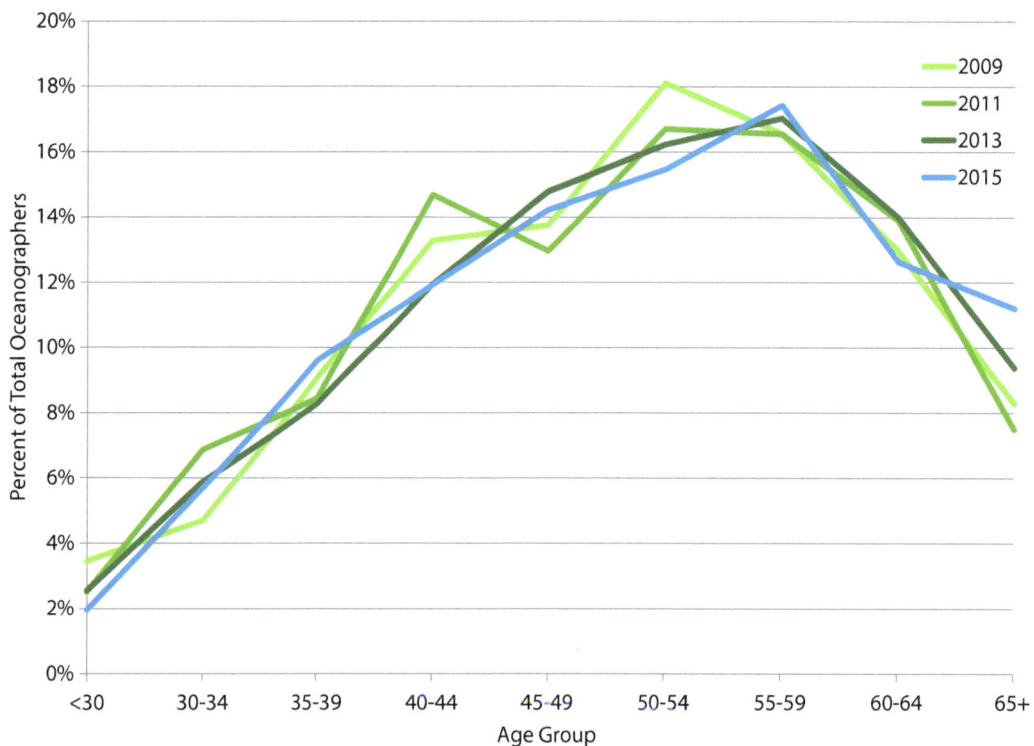

AGI Geoscience Workforce Program; Data derived from the Office of Personnel Management fedscope database

Figure 4.30: Age Distribution of Geoscientists in the U.S. Geological Survey

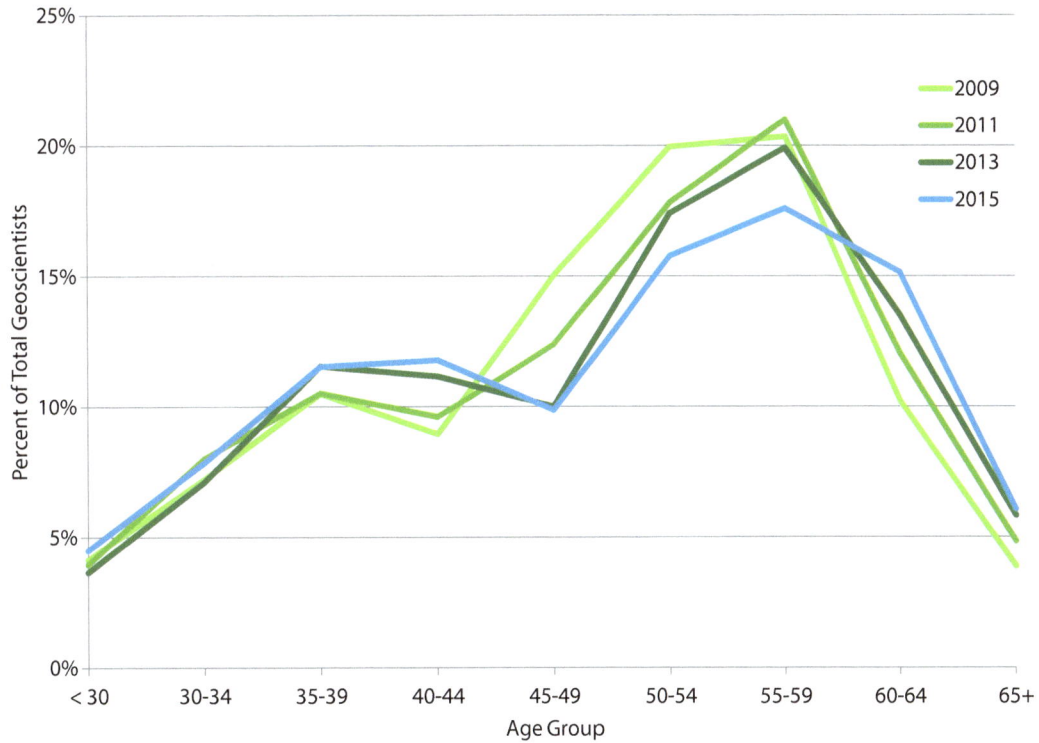

AGI Geoscience Workforce Program; Data derived from the Office of Personnel Management fedscope database

Figure 4.31: Oil and Gas Industry Supply and Demand for Geoscientists

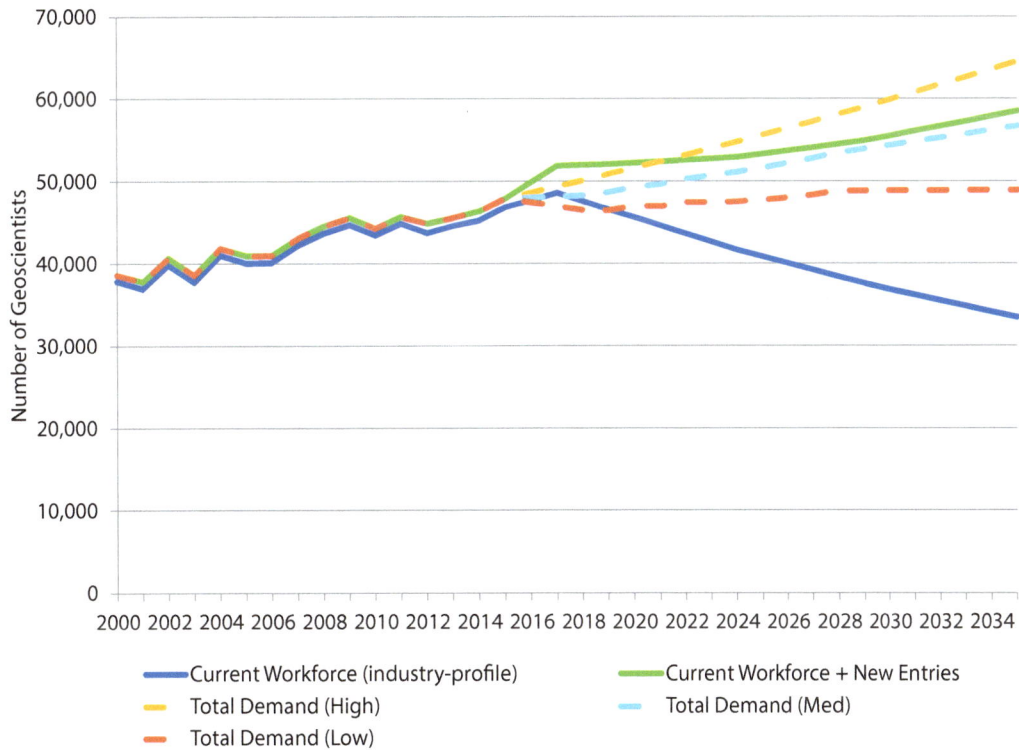

AGI Geoscience Workforce Program

Figure 4.32: Age Distribution of Geoscientists in Mining, 2014

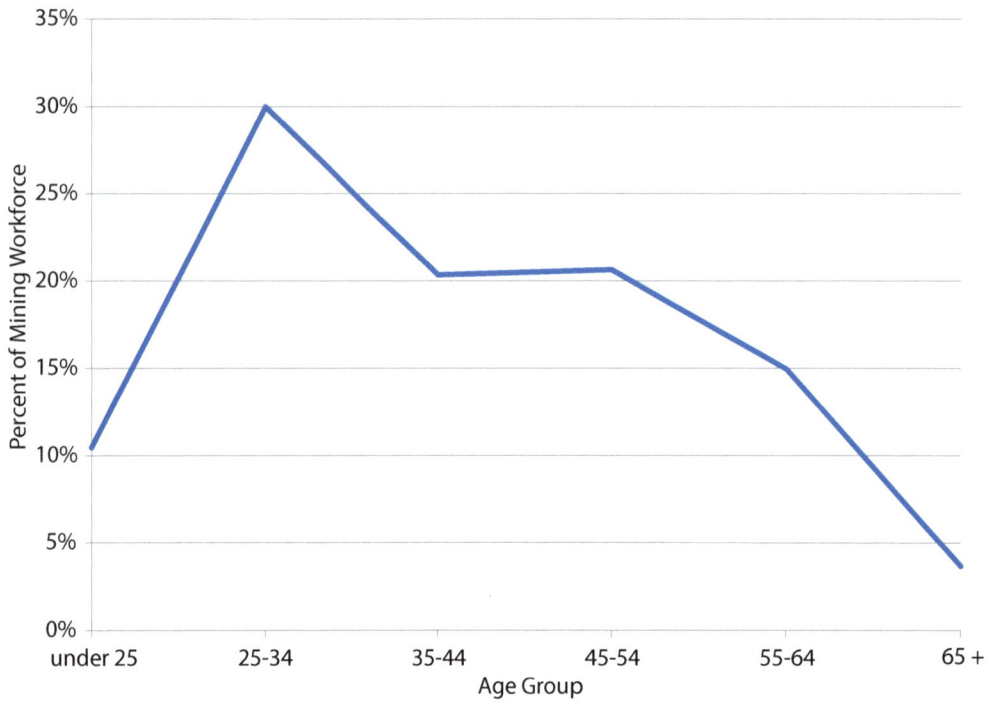

AGI Geoscience Workforce Program; Data provided by the National Mining Association

Figure 4.33: Age Distribution of Geoscientists in Support Activities for Mining and Oil & Gas, 2014

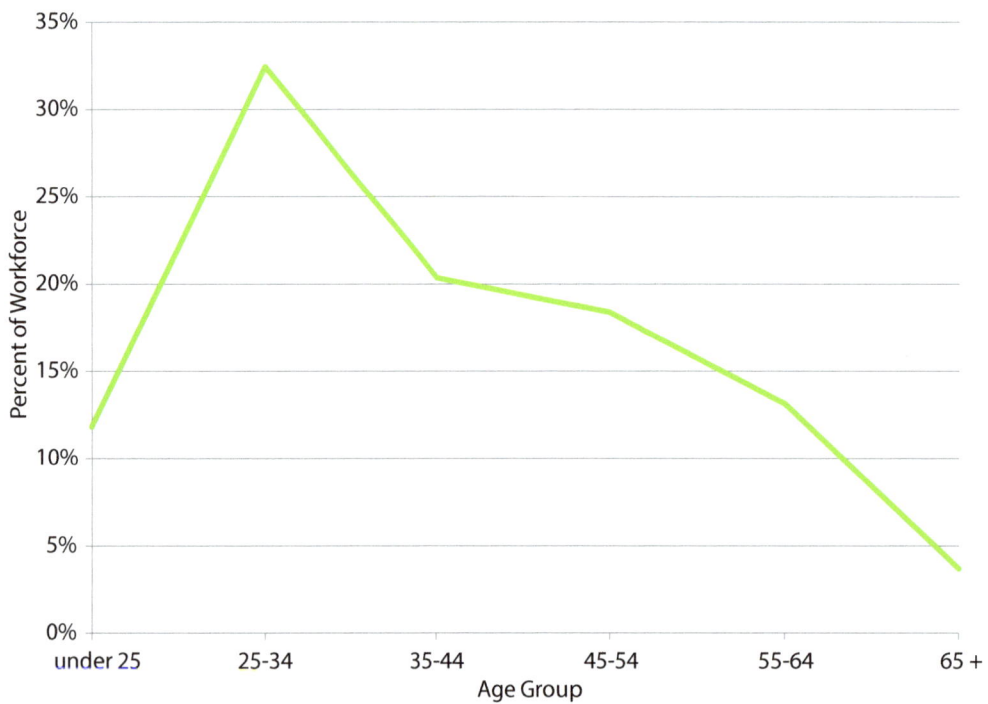

AGI Geoscience Workforce Program; Data provided by the National Mining Association

Figure 4.34: Trends in Faculty Rank Distribution at Four-Year Institutions (1980-2015)

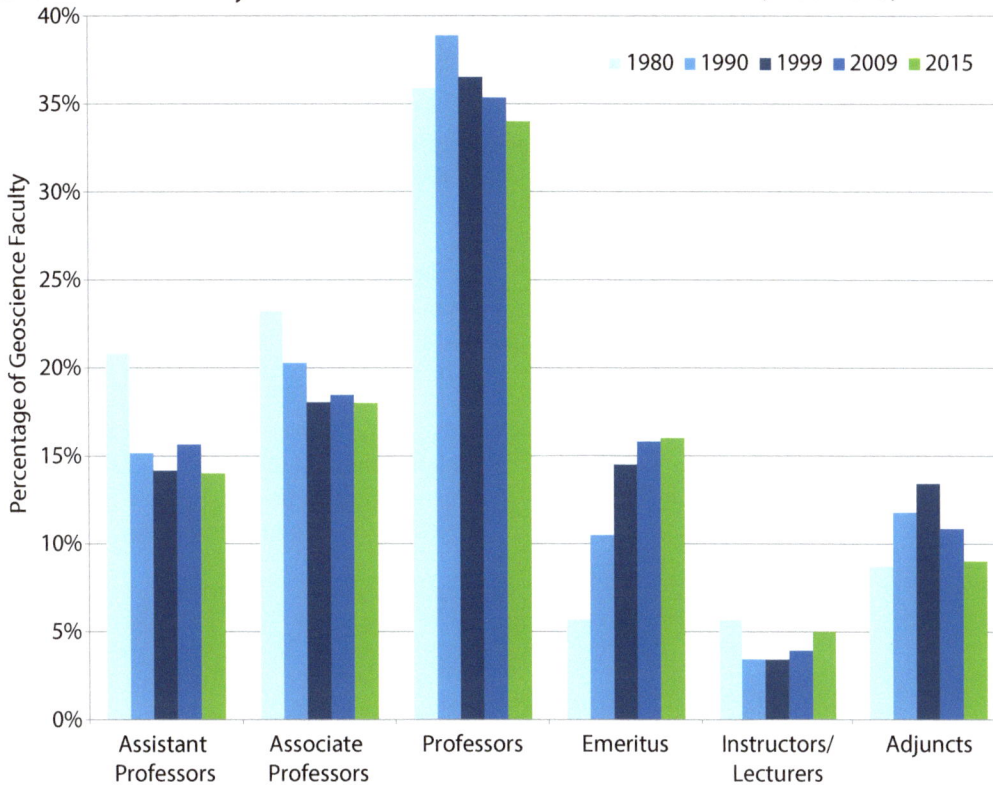

AGI Geoscience Workforce Program; Data derived from AGI's Directory of Geoscience Departments database

Figure 4.35: Age Distribution of Geoscience Faculty Members

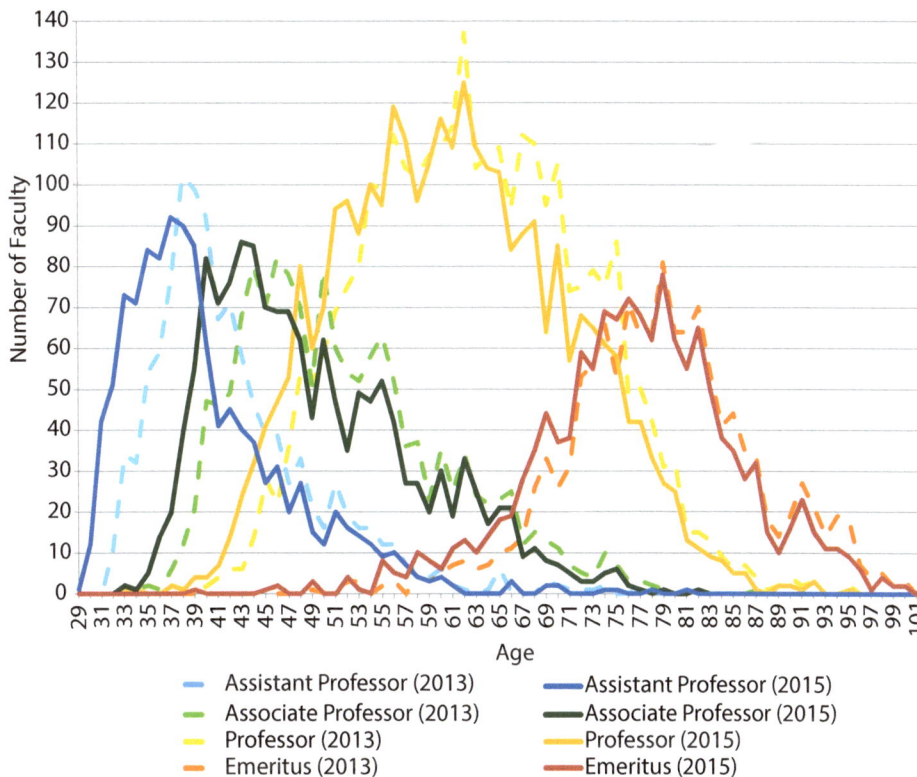

AGI Geoscience Workforce Program; Data derived from AGI's Directory of Geoscience Departments database

Geoscience Employment Projections 2014–2024

According to the U.S. Bureau of Labor Statistics, there were a total of 324,411 geoscientist jobs in 2014, and they are projecting job growth to 355,862 geoscientist jobs in 2024—a 10% increase between 2014 and 2024 (Figure 4.36). This growth translates in all major industries hiring geoscientists except for the mining (except oil and gas) industry and the federal government (Figure 4.33). The U.S. Bureau of Labor Statistics is projecting a 6% loss in jobs in the mining industry and an 8% loss of jobs in the federal government for geoscientists. Environmental engineers and environmental scientists are projected to have the highest percent change in number of available jobs by 2024. In the 2014 report, a high percentage of change was predicted among petroleum engineers at and increase of 26%, but the employment and economic changes in the petroleum industry has changed the updated decade projections provided by the BLS changing the prediction to an increase of 10%. The 2014 report also displayed a high percentage of change among geographers between 2012 and 2022 at an increase of 29%, but the updated projections for 2014-2024 show no growth in this occupation.

Table 4.1 documents the number of jobs in 2014 and projected for 2024 for different occupations within various industries known for hiring geoscientists, as well as the 2014 median annual salary for these occupations.

Figure 4.36: Employment Projection for Geoscience Occupations (2014-2024)

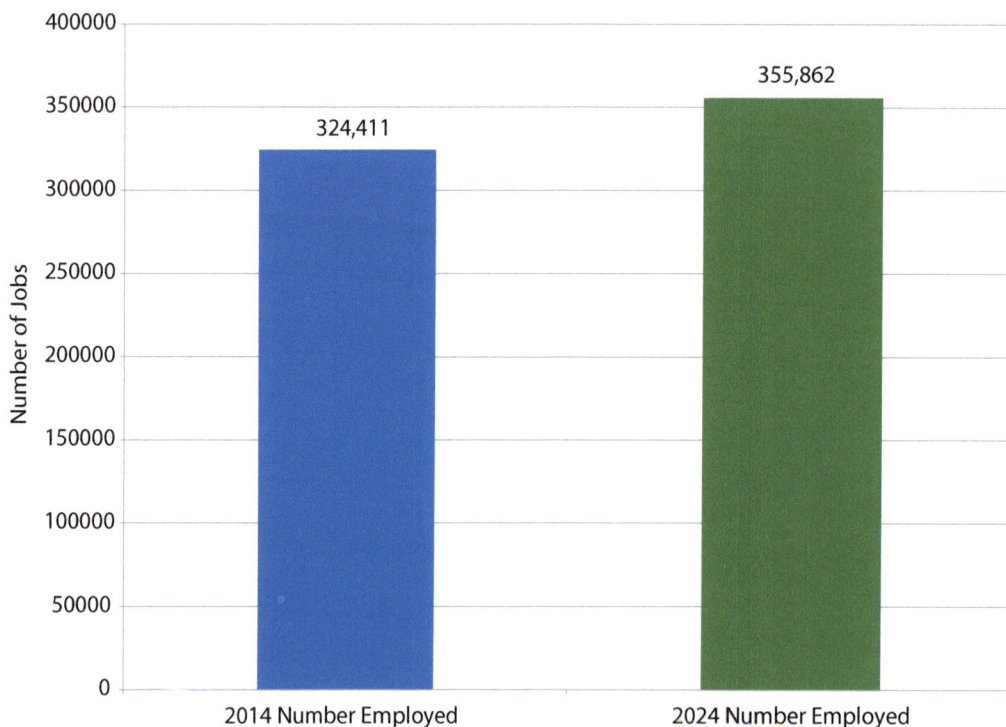

AGI Geoscience Workforce Program; Data derived from the U.S. Bureau of Labor Statistics, Employment Projections

Figure 4.37: Employment Projections for All Geoscience Occupations by Industry Sector (2014-2024)

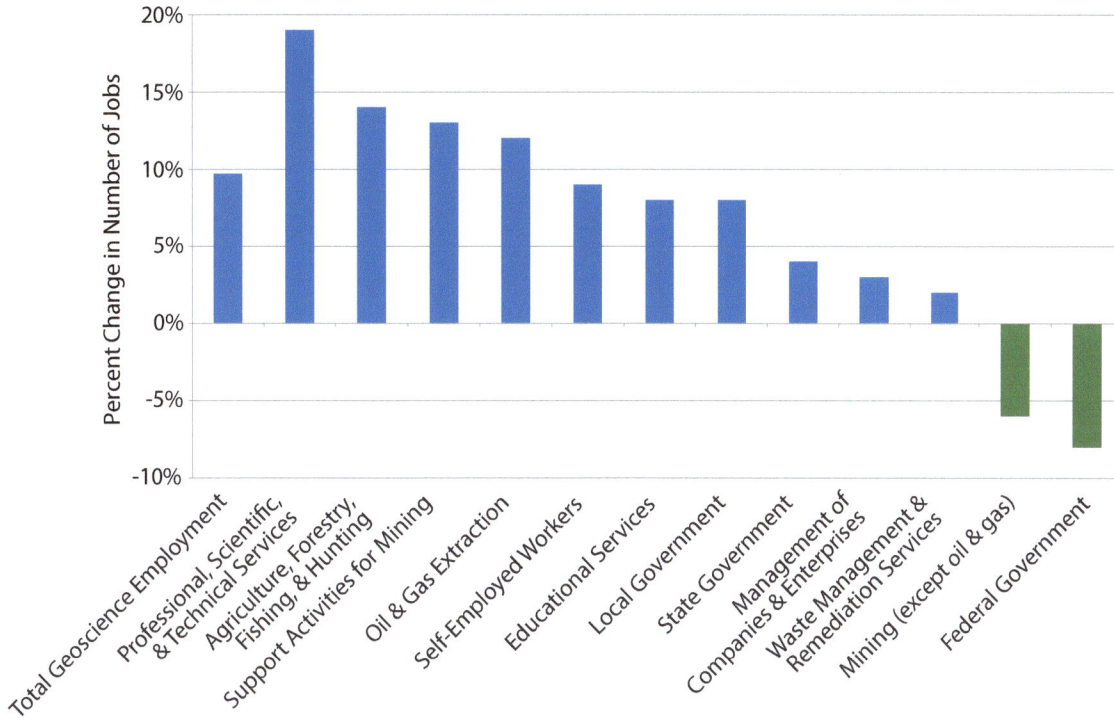

AGI Geoscience Workforce Program; Data derived from the U.S. Bureau of Labor Statistics, Employment Projections

Figure 4.38: Employment Projections for Detailed Geoscience Occupations (2014-2024)

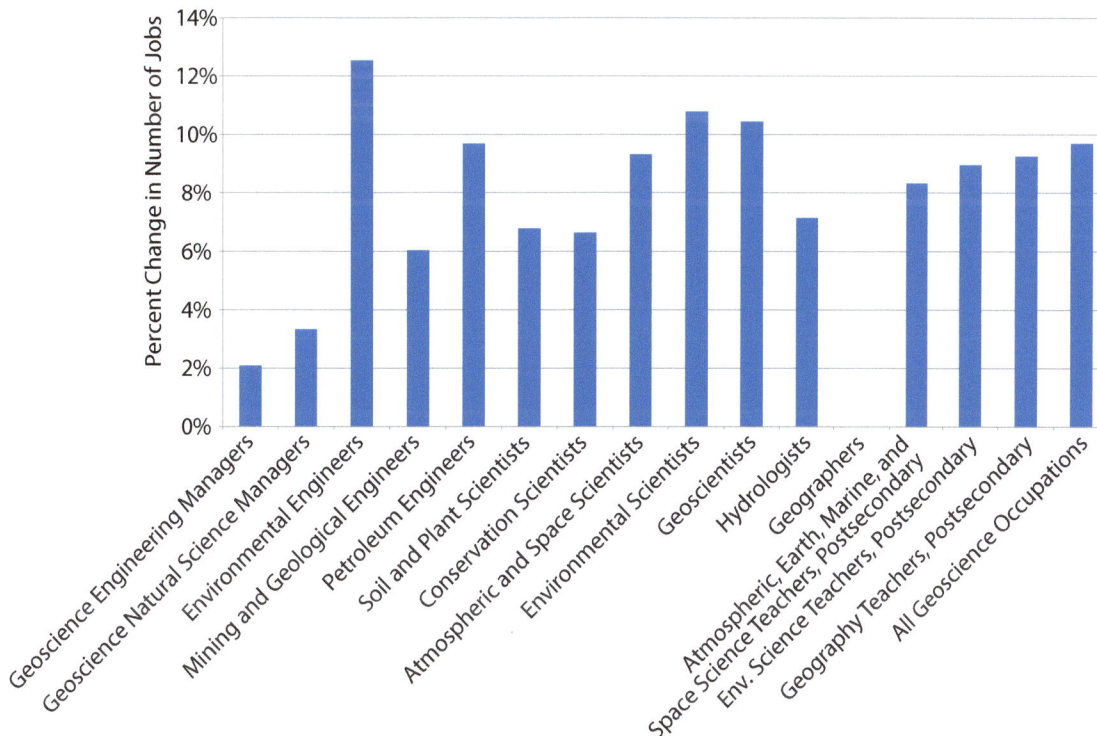

AGI Geoscience Workforce Program; Data derived from the U.S. Bureau of Labor Statistics, Employment Projections

Table 4.1: U.S. Bureau of Labor Statistics Current and Projected Geoscience Employment

Occupation	2014 Number Employed	2024 Number Employed	2014-2024 Percent Change	2014 Median Annual Salary (OES)
Total Employment, All Workers				
Engineering managers (*)	8,146	8,315	2%	-
Natural science managers (*)	2,465	2,546	3%	$120,050
Environmental engineers	55,100	62,000	13%	$83,360
Mining and geological engineers, including mining safety engineers	8,300	8,800	6%	$90,160
Petroleum engineers	35,100	38,500	10%	$130,050
Soil and plant scientists	17,700	18,900	7%	$59,920
Conservation scientists	21,100	22,500	7%	$61,860
Atmospheric and space scientists	11,800	12,900	9%	$87,980
Environmental scientists	94,600	104,800	11%	$66,250
Geoscientists, except hydrologists and geographers	36,400	40,200	10%	$89,910
Hydrologists	7,000	7,500	7%	$78,370
Geographers	1,400	1,400	0%	$76,420
Atmospheric, earth, marine, and space science teachers, postsecondary	13,200	14,300	8%	$81,780
Environmental science teachers, postsecondary	6,700	7,300	9%	$77,470
Geography teachers, postsecondary	5,400	5,900	9%	$71,320
All Geoscience Occupations	324,411	355,862	10%	
Self-Employed Workers, All Jobs				
Engineering managers (*)	4	6	39%	-
Environmental engineers	400	500	25%	-
Petroleum engineers	600	700	17%	-
Soil and plant scientists	600	700	17%	-
Conservation scientists	300	300	0%	-
Atmospheric and space scientists	200	200	0%	-
Environmental scientists	1,600	1,700	6%	-
Geoscientists, except hydrologists and geographers	900	900	0%	-
Hydrologists	100	100	0%	-
All Geoscience Occupations	4,704	5,106	9%	
Agriculture, Forestry, Fishing, and Hunting				
Natural science managers (*)	0	0	0%	$104,080
Soil and plant scientists	1,500	1,600	7%	$60,020
Conservation scientists	200	300	50%	$64,220
Environmental scientists	300	400	33%	-
Geoscientists, except hydrologists and geographers	200	200	0%	-
All Geoscience Occupations	2,200	2,500	14%	
Oil and Gas Extraction				
Engineering managers (*)	1,507	1,663	10%	-
Natural science managers (*)	112	111	-1%	$170,660
Environmental engineers	500	600	20%	$105,630
Mining and geological engineers, including mining safety engineers	900	1,000	11%	$132,660
Petroleum engineers	15,800	17,700	12%	$140,910
Environmental scientists	1,200	1,300	8%	$91,630
Geoscientists, except hydrologists and geographers	8,000	9,000	13%	$129,550
All Geoscience Occupations	28,019	31,374	12%	

Occupation	2014 Number Employed	2024 Number Employed	2014-2024 Percent Change	2014 Median Annual Salary (OES)
Mining (Except Oil and Gas)				
Engineering managers (*)	101	99	-3%	-
Environmental engineers	400	300	-25%	$85,450
Mining and geological engineers, including mining safety engineers	2,500	2,400	-4%	$86,910
Environmental scientists	100	100	0%	$71,910
Geoscientists, except hydrologists and geographers	500	500	0%	$94,180
All Geoscience Occupations	3,601	3,399	-6%	
Support Activities for Mining				
Engineering managers (*)	225	257	14%	-
Environmental engineers	100	100	0%	$106,390
Mining and geological engineers, including mining safety engineers	900	1,000	11%	$117,880
Petroleum engineers	5,600	6,400	14%	$106,120
Environmental scientists	100	100	0%	$68,600
Geoscientists, except hydrologists and geographers	1,600	1,800	13%	$109,120
All Geoscience Occupations	8,525	9,657	13%	
Utilities				
Engineering managers (*)	242	228	-6%	-
Natural science managers (*)	6	6	4%	$118,520
Environmental engineers	800	800	0%	$82,720
Mining and geological engineers, including mining safety engineers	0	0	0%	$82,570
Petroleum engineers	1000	900	-10%	$105,470
Conservation scientists	100	100	-7%	$77,350
Environmental scientists	1,400	1,300	-7%	$89,890
Geoscientists, except hydrologists and geographers	200	200	0%	$82,640
Hydrologists	100	100	0%	$89,860
All Geoscience Occupations	3,848	3,634	-6%	
Construction				
Engineering managers (*)	22	22	-2%	-
Environmental engineers	200	200	0%	$83,910
Environmental scientists	100	100	0%	$70,250
Geoscientists, except hydrologists and geographers	0	0	0%	$66,370
All Geoscience Occupations	322	322	0%	
Manufacturing				
Engineering managers (*)	581	546	-6%	-
Natural science managers (*)	61	60	0%	$127,000
Environmental engineers	3,000	2,800	-7%	$92,830
Mining and geological engineers, including mining safety engineers	100	100	0%	$86,910
Petroleum engineers	3,300	3,200	-3%	$123,220
Soil and plant scientists	200	100	-50%	$68,140
Atmospheric and space scientists	0	0	0%	$115,150
Environmental scientists	1,600	1,600	0%	$80,810
Geoscientists, except hydrologists and geographers	300	300	0%	$104,690
All Geoscience Occupations	9,142	8,707	-5%	

Occupation	2014 Number Employed	2024 Number Employed	2014-2024 Percent Change	2014 Median Annual Salary (OES)
Wholesale Trade				
Engineering managers (*)	51	48	-5%	-
Natural science managers (*)	12	10	-17%	$136,000
Environmental engineers	300	300	0%	$91,510
Petroleum engineers	100	100	0%	$118,920
Soil and plant scientists	2,100	2,000	-5%	$60, 120
Environmental scientists	100	100	0%	$65,460
All Geoscience Occupations	2,663	2,559	-4%	
Transportation and Warehousing				
Engineering managers (*)	66	67	1%	-
Environmental engineers	200	200	0%	$94,270
Petroleum engineers	1,300	1,400	0%	$118,270
Soil and plant scientists	0	0	0%	$62,390
Atmospheric and space scientists	100	100	0%	$71,100
Environmental scientists	500	500	-33%	$105,980
Geoscientists, except hydrologists and geographers	0	0	0%	$72,410
All Geoscience Occupations	2,166	2,267	5%	
Information				
Engineering managers (*)	5	4	-13%	-
Atmospheric and space scientists	800	800	0%	$88,740
All Geoscience Occupations	805	804	0%	
Finance and Insurance				
Natural science managers (*)	0	0	-14%	$116,150
Petroleum engineers	100	100	0%	-
All Geoscience Occupations	100	100	0%	
Professional, Scientific, and Technical Services				
Engineering managers (*)	2,938	3,296	12%	-
Natural science managers (*)	1,012	1,114	10%	$138,100
Environmental engineers	28,900	35,700	24%	$83,680
Mining and geological engineers, including mining safety engineers	2,600	3,000	15%	$82,030
Petroleum engineers	3,400	3,900	15%	$149,500
Soil and plant scientists	6,000	6,900	15%	$61,660
Conservation scientists	1,300	1,500	15%	$66,370
Atmospheric and space scientists	4,800	5,800	21%	$87,370
Environmental scientists	37,800	45,300	20%	$68,050
Geoscientists, except hydrologists and geographers	14,600	17,000	16%	$78,460
Hydrologists	2,900	3,500	21%	$87,030
Geographers	300	300	0%	$61,910
All Geoscience Occupations	106,550	127,311	19%	

Occupation	2014 Number Employed	2024 Number Employed	2014-2024 Percent Change	2014 Median Annual Salary (OES)
Architectural, Engineering, and Related Services				
Engineering managers (*)	4,022	4,521	12%	-
Natural science managers (*)	238	261	10%	$107,770
Environmental engineers	16,200	19,700	22%	$84,870
Mining and geological engineers, including mining safety engineers	2,300	2,500	9%	$79,510
Petroleum engineers	1,600	1,800	13%	$153,000
Soil and plant scientists	400	400	0%	$51,670
Conservation scientists	300	400	33%	$67,440
Atmospheric and space scientists	200	200	0%	$91,880
Environmental scientists	12,700	14,000	10%	$64,220
Geoscientists, except hydrologists and geographers	8,300	9,200	11%	$81,480
Hydrologists	1,200	1,400	17%	$88,590
Geographers	200	200	0%	$57,890
All Geoscience Occupations	**47,660**	**54,582**	**15%**	
Testing Laboratories				
Engineering managers (*)	144	151	5%	-
Natural science managers (*)	98	105	8%	$98,590
Environmental engineers	700	800	14%	$77,910
Mining and geological engineers, including mining safety engineers	100	100	0%	$69,770
Petroleum engineers	0	0	0%	$94,730
Soil and plant scientists	400	400	0%	$50,410
Environmental scientists	2,900	3,100	7%	$52,450
Geoscientists, except hydrologists and geographers	500	500	0%	$60,730
All Geoscience Occupations	**4,841**	**5,157**	**7%**	
Computer Systems Design and Related Services				
Engineering managers (*)	3	3	-3%	-
Natural science managers (*)	0	0	0%	$132,950
Environmental engineers	0	0	0%	$92,850
Atmospheric and space scientists	200	200	0%	$113,220
Environmental scientists	300	300	0%	$58,250
Geoscientists, except hydrologists and geographers	200	200	0%	$123,430
All Geoscience Occupations	**703**	**703**	**0%**	
Management, Scientific, and Technical Consulting Services				
Engineering managers (*)	791	992	25%	-
Natural science managers (*)	598	744	24%	$125,640
Environmental engineers	11,200	14,300	28%	$79,970
Mining and geological engineers, including mining safety engineers	300	400	33%	$135,430
Petroleum engineers	1,100	1,400	27%	$135,530
Soil and plant scientists	2,300	3,000	30%	$56,130
Conservation scientists	600	800	33%	$63,660
Atmospheric and space scientists	400	500	25%	$67,560
Environmental scientists	21,700	27,600	27%	$68,410
Geoscientists, except hydrologists and geographers	5,300	6,800	28%	$73,840
Hydrologists	1,600	2,000	25%	$85,810
All Geoscience Occupations	**45,990**	**58,635**	**27%**	

Occupation	2014 Number Employed	2024 Number Employed	2014-2024 Percent Change	2014 Median Annual Salary (OES)
Scientific Research and Development Services				
Engineering managers (*)	302	308	2%	-
Natural science managers (*)	565	573	1%	$145,630
Environmental engineers	1,300	1,300	0%	$96,050
Mining and geological engineers, including mining safety engineers	0	0	0%	$57,640
Petroleum engineers	600	600	0%	$143,300
Soil and plant scientists	3,200	3,400	6%	$67,780
Conservation scientists	300	300	0%	$75,170
Atmospheric and space scientists	2,500	2,600	4%	$93,600
Environmental scientists	3,100	3,200	3%	$86,980
Geoscientists, except hydrologists and geographers	600	600	0%	$84,660
Hydrologists	100	100	0%	$88,700
Geographers	0	0	0%	$65,650
All Geoscience Occupations	**12,566**	**12,981**	**3%**	
Other Professional, Scientific, and Technical Services				
Engineering managers (*)	10	11	12%	-
Natural science managers (*)	5	11	125%	$120,310
Environmental engineers	0	0	0%	$74,260
Atmospheric and space scientists	1,500	2,200	47%	$67,450
Environmental scientists	100	100	0%	$51,480
Geoscientists, except hydrologists and geographers	200	200	0%	$59,090
All Geoscience Occupations	**1,814**	**2,522**	**39%**	
Management of Companies and Enterprises				
Engineering managers (*)	295	300	2%	-
Natural science managers (*)	116	116	0%	$160,200
Environmental engineers	2,900	2,900	0%	$94,930
Mining and geological engineers, including mining safety engineers	400	400	0%	$105,040
Petroleum engineers	3,200	3,300	3%	$152,450
Soil and plant scientists	300	400	33%	$79,270
Atmospheric and space scientists	0	0	0%	$97,160
Environmental scientists	900	900	0%	$100,130
Geoscientists, except hydrologists and geographers	1,400	1,500	7%	$149,010
All Geoscience Occupations	**9,511**	**9,816**	**3%**	
Administrative and Support and Waste Management and Remediation Services				
Engineering managers (*)	34	34	1%	-
Natural science managers (*)	5	4	-11%	$118,780
Environmental engineers	2,900	2,900	0%	$83,100
Mining and geological engineers, including mining safety engineers	100	100	0%	$124,510
Petroleum engineers	100	100	0%	$139,020
Soil and plant scientists	300	300	0%	$50,240
Conservation scientists	100	100	0%	$70,270
Environmental scientists	2,200	2,200	0%	$68,880
Geoscientists, except hydrologists and geographers	300	400	33%	$86,470
All Geoscience Occupations	**6,039**	**6,138**	**2%**	

Occupation	2014 Number Employed	2024 Number Employed	2014-2024 Percent Change	2014 Median Annual Salary (OES)
Waste Management and Remediation Services				
Engineering managers (*)	168	178	6%	-
Natural science managers (*)	34	36	6%	$114,310
Environmental engineers	2,500	2,500	0%	$82,000
Environmental scientists	1,500	1,500	0%	$63,440
Geoscientists, except hydrologists and geographers	200	200	0%	$80,940
All Geoscience Occupations	**4,402**	**4,414**	**0%**	
Educational Services, Public and Private				
Engineering managers (*)	80	80	1%	-
Natural science managers (*)	200	217	9%	$98,020
Environmental engineers	200	200	0%	$74,160
Petroleum engineers	0	0	0%	$55,920
Soil and plant scientists	3,200	3,300	3%	$51,770
Conservation scientists	1,000	1,100	10%	$58,390
Atmospheric and space scientists	2,400	2,800	17%	$69,980
Environmental scientists	3,800	4,000	5%	$61,810
Geoscientists, except hydrologists and geographers	2,600	2,700	4%	$66,090
Geographers	100	100	0%	$41,940
Atmospheric, earth, marine, and space science teachers, postsecondary	13,200	14,300	8%	$83,130
Environmental science teachers, postsecondary	6,700	7,300	9%	$78,770
Geography teachers, postsecondary	5,400	5,900	9%	$75,400
All Geoscience Occupations	**38,880**	**41,998**	**8%**	
Junior Colleges; State, Local, and Private				
Atmospheric, earth, marine, and space science teachers, postsecondary	2,900	3,200	10%	$69,200
Environmental science teachers, postsecondary	800	1,000	25%	$61,820
Geography teachers, postsecondary	1,000	1,100	10%	$72,550
All Geoscience Occupations	**4,700**	**5,300**	**13%**	
Colleges, Universities, and Professional Schools; State, Local, and Private				
Engineering managers (*)	92	101	9%	-
Natural science managers (*)	246	263	7%	$97,780
Environmental engineers	200	200	0%	$73,850
Petroleum engineers	0	0	0%	$55,920
Soil and plant scientists	3,200	3,300	3%	$51,760
Conservation scientists	900	1,000	11%	$58,100
Atmospheric and space scientists	2,400	2,700	13%	$71,010
Environmental scientists	3,700	3,900	5%	$61,160
Geoscientists, except hydrologists and geographers	2,600	2,700	4%	$66,230
Hydrologists	100	100	0%	-
Geographers	100	100	0%	$41,160
Atmospheric, earth, marine, and space science teachers, postsecondary	10,500	11,000	5%	$88,370
Environmental science teachers, postsecondary	5,800	6,300	9%	$81,680
Geography teachers, postsecondary	4,400	4,700	7%	$75,870
All Geoscience Occupations	**34,239**	**36,364**	**6%**	

Occupation	2014 Number Employed	2024 Number Employed	2014-2024 Percent Change	2014 Median Annual Salary (OES)
Federal Government, Excluding Postal Service				
Engineering managers (*)	660	606	-8%	-
Natural science managers (*)	804	736	-8%	$110,900
Environmental engineer	3,500	3,100	-11%	$101,640
Mining and geological engineers, including mining safety engineers	100	100	0%	$88,910
Petroleum engineers	300	300	0%	$99,910
Soil and plant scientists	1,500	1,300	-13%	$74,380
Conservation scientists	7,200	7,000	-3%	$74,180
Atmospheric and space scientists	3,100	2,800	-10%	$99,920
Environmental scientists	5,800	5,200	-10%	$99,260
Geoscientists, except hydrologists and geographers	2,400	2,200	-8%	$96,480
Hydrologists	1,900	1,800	-5%	$85,830
Geographers	800	700	-13%	$83,200
All Geoscience Occupations	**28,065**	**25,842**	**-8%**	
State Government, Excluding Education and Hospitals				
Engineering managers (*)	437	443	1%	-
Natural science managers (*)	619	646	4%	$77,290
Environmental engineers	8,400	8,600	2%	$75,440
Mining and geological engineers, including mining safety engineers	500	500	0%	$93,860
Petroleum engineers	100	100	0%	$114,020
Soil and plant scientists	500	500	0%	$50,090
Conservation scientists	5,100	5,500	8%	$53,410
Atmospheric and space scientists	300	300	0%	$66,060
Environmental scientists	20,700	21,800	5%	$60,280
Geoscientists, except hydrologists and geographers	2,800	2,900	4%	$69,790
Hydrologists	1,200	1,200	0%	$62,600
Geographers	100	100	0%	$57,920
All Geoscience Occupations	**40,755**	**42,589**	**4%**	
Local Government, Excluding Education and Hospitals				
Engineering managers (*)	681	722	6%	-
Natural science managers (*)	205	219	6%	$100,150
Environmental engineers	3,800	4,000	6%	$80,100
Mining and geological engineers, including mining safety engineers	500	500	0%	$75,220
Soil and plant scientists	900	900	0%	$53,880
Conservation scientists	3,700	4,100	11%	$48,040
Atmospheric and space scientists	100	100	0%	$102,200
Environmental scientists	12,700	13,900	9%	$65,320
Geoscientists, except hydrologists and geographers	200	200	0%	$88,170
Hydrologists	600	600	0%	$71,670
All Geoscience Occupations	**23,386**	**25,241**	**8%**	

(*): Engineering managers and Natural science manager employment numbers were estimated from the federal data by dividing the total non-manager geoscientists by the total number of non-manager S&E employees per industry and then multiplying this result by the total number of engineering (or natural science) managers per industry.

AGI Geoscience Workforce Program; Data derived from the U.S. Bureau of Labor Statistics Employment Projections

Chapter 5: Trends in Economic Metrics and Drivers of the Geoscience Workforce

Tracking the economic metrics related to the geoscience workforce can indicate potential changes in the supply and demand for the future workforce, which in turn impacts students finishing geoscience degrees at four-year universities. Therefore, this chapter looks at the changes in federal funding for geoscience research, economic metrics in the petroleum and mining industries, and the contribution of geoscience to the overall U.S. gross domestic product.

The percentage of federal research funding awarded to geoscience research declined over the years from 13% in 1970 to 7% in 2015, but the total dollar amount of federal geoscience research funding has increased steadily to $4 billion in 2013. However, the difference between the amount spent on applied compared to basic research has increased with approximately 2% more of the research funds spent on basic research from 2013-2015. This increase in spending for basic research can be attributed to an increase in the funding for basic research in the atmospheric sciences.

This report clearly shows the economic downturn in the oil and gas industry starting at the end of 2014. The price of oil, the gross domestic product contributed by geoscientists in the oil and gas industry, and number of working oil rigs show a rapid decline over the past couple of years. At the end of 2015, the United States was operating 36% of the drilling rigs in the world compared to 52% at the height of the industry in 2014, and this downturn has affected all areas of the industry with decreases in the number of land rigs, off-shore rigs, crude oil rigs, and natural gas rigs.

Since the economic recession in the United States in 2009, the U.S. non-fuel mines yielding metal ore and industrial minerals have seen an increase in the material handled by these mines, as well as the value of these yields.

All major employment sectors that directly hire geoscientists currently make up approximately 0.52% of the total U.S. economy equaling $91 billion of the U.S. gross domestic product. This is a decrease of approximately $9 billion due to the economic downturn in the oil and gas industry. The projected GDP contributed by geoscientists in the oil and gas industry in 2024 is approximately $51 billion, which is $2 billion less than the contributed GDP by geoscientists in this industry in 2012. However, this projection will change if there is a strong bounce back for the U.S. oil and gas industry in the next few years.

Federal Research Funding for the Geosciences

There was an overall decrease in federal funding for applied and basic research in the geosciences since 1996 reaching a low of 6% for basic research and 5% for applied research in 2012, but from 2012-2015 there was a slight increase in federal funding for basic research in the geosciences to 8% (Figure 5.1). The overall amount of money spent on geoscience research by the federal government has been fluctuating between $3-3.7 billion, but in

2014, total federal funds spent on the geosciences reached $4 billion, which is higher than the amount spent in 2009 during the ARRA stimulus funding increase (Figure 5.2). Since 2010, there has been an increase in federal funding for the atmospheric sciences and a decrease in federal funding for environmental sciences. This shift can be attributed to an increase in basic research funding for the atmospheric sciences from 2009-2014 (Figure 5.3).

Figure 5.1: Percentage of Federal Research Funding Applied to the Geosciences

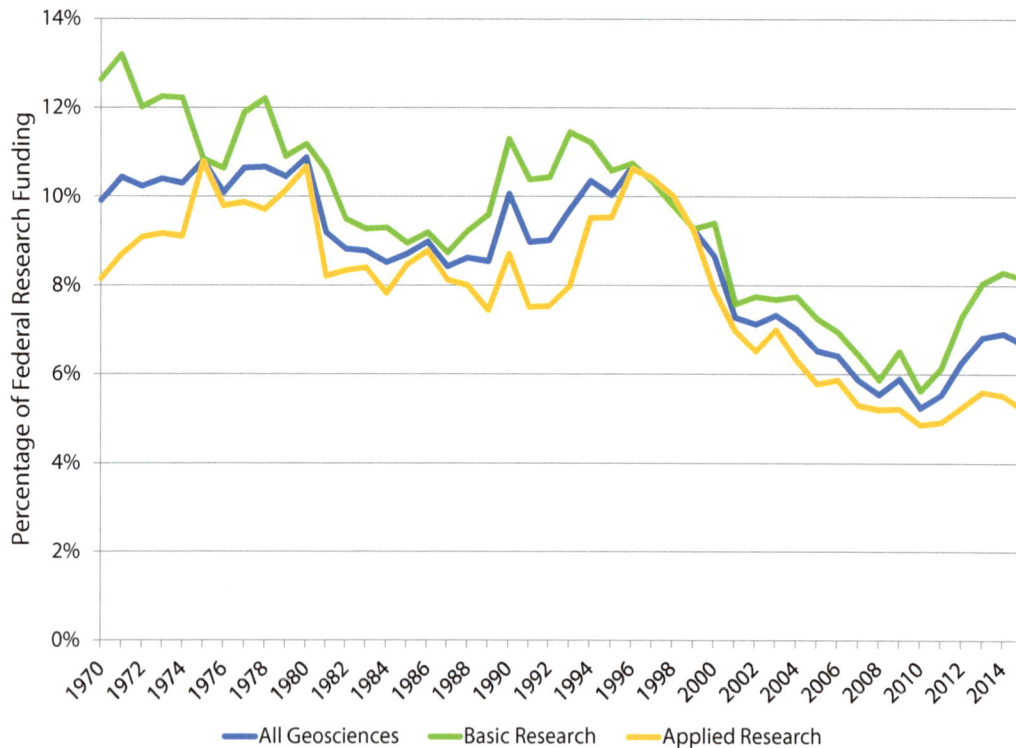

AGI Geoscience Workforce Program; Data derived from NSF/SRS Survey of Federal Funds for Research & Development

Figure 5.2: Total Federal Research Funding of the Geosciences

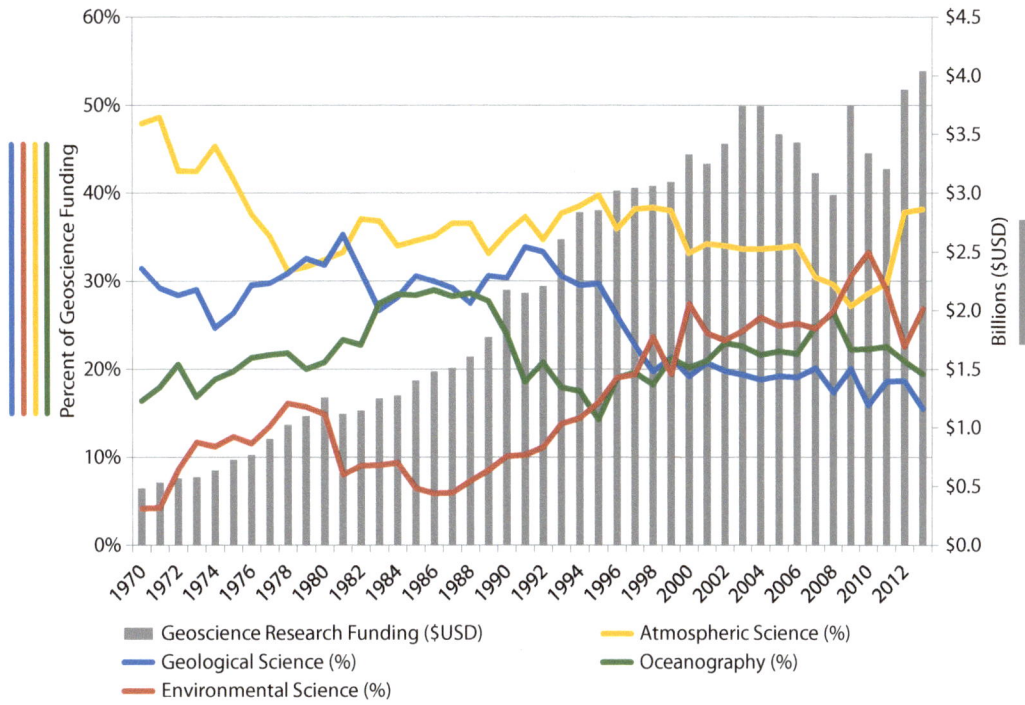

AGI Geoscience Workforce Program; Data derived from the NSF/SRS Survey of Federal Funds for Research & Development

Figure 5.3: Federal Funding of Basic Research in the Geosciences

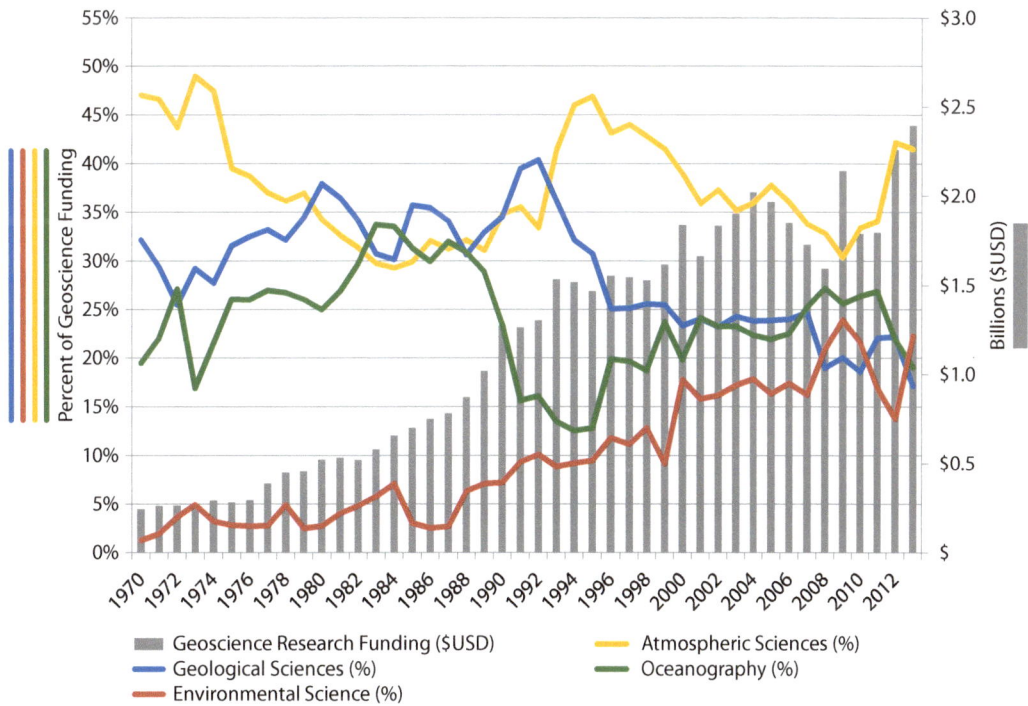

AGI Geoscience Workforce Program; Data derived from NSF/SRS Survey of Federal Funds for Research & Development

Figure 5.4: Federal Funding of Applied Research in the Geosciences

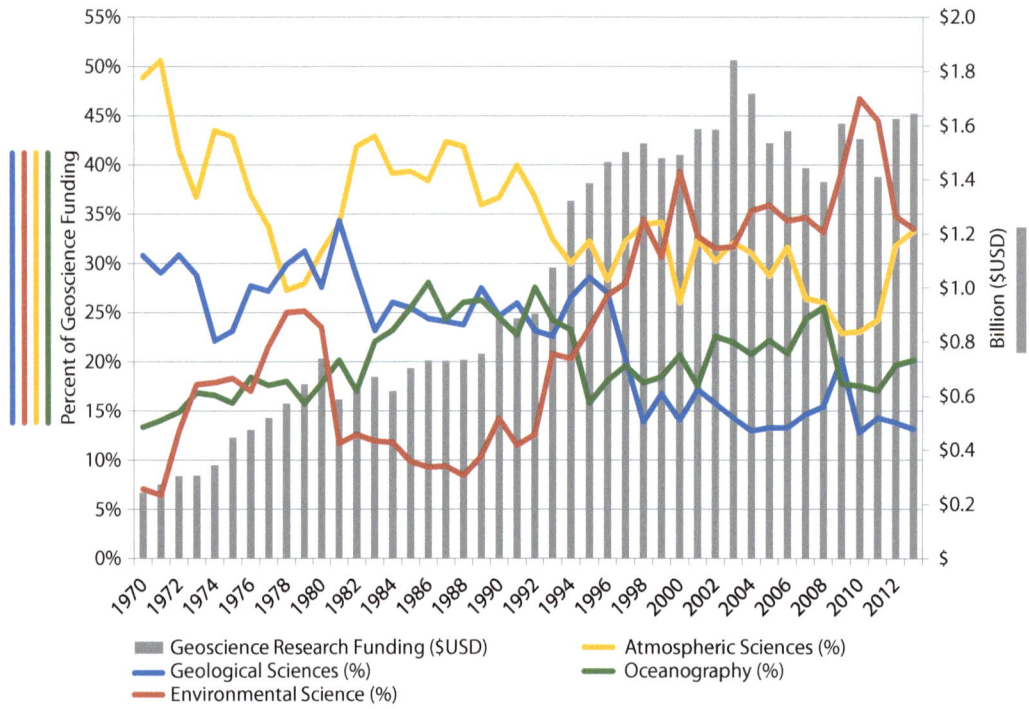

Geoscience Research Funding ($USD)   Atmospheric Sciences (%)
Geological Sciences (%)   Oceanography (%)
Environmental Science (%)

AGI Geoscience Workforce Program; Data derived from NSF/SRS Survey of Federal Funds for Research & Development

Commodity Prices and Output

Figure 5.5 shows the changes in spot prices of crude oil per barrel (BBL) in US dollars and in euros from 2001 to 2016, which covers the recent economic recession, subsequent recovery period, and recent economic downturn in the petroleum industry. The graph also shows the number of barrels per troy ounce of gold. The price of oil, relative to the dollar, grew faster than it did relative to the euro through the recession and this trend has continued through the economic recovery to the present. Some amount of the rise in oil prices can be attributed to the fall of the value of the dollar. However, in 2014, the price per barrel with the dollar and the euro dramatically decreased due to the economic downturn within the petroleum industry.

The total domestic commodity output data for the petroleum and mining industries shows a steady increase from 2002 to 2008, followed by a sharp decline in 2009 due to the economic recession (Figure 5.6). However, the industries were able to bounce back quickly during the economic recovery and into 2014. Because the data is only up to date to 2014, the economic downturn in the petroleum industry is not evident in this graph or in the data for the gross operating surplus.

The gross operating surplus for the petroleum and mining industries grew steadily until 2008, after which only mining continued to increase (Figure 5.7). After the recession, the oil and gas extraction industry rebounded and continued to increase into 2014. The gross operating surplus for the mining industry increased from 2009-2012, but the gross operating surplus has remained steady at approximately $50 billion through 2014.

Figure 5.5: Price of Oil by Currency and by Gold

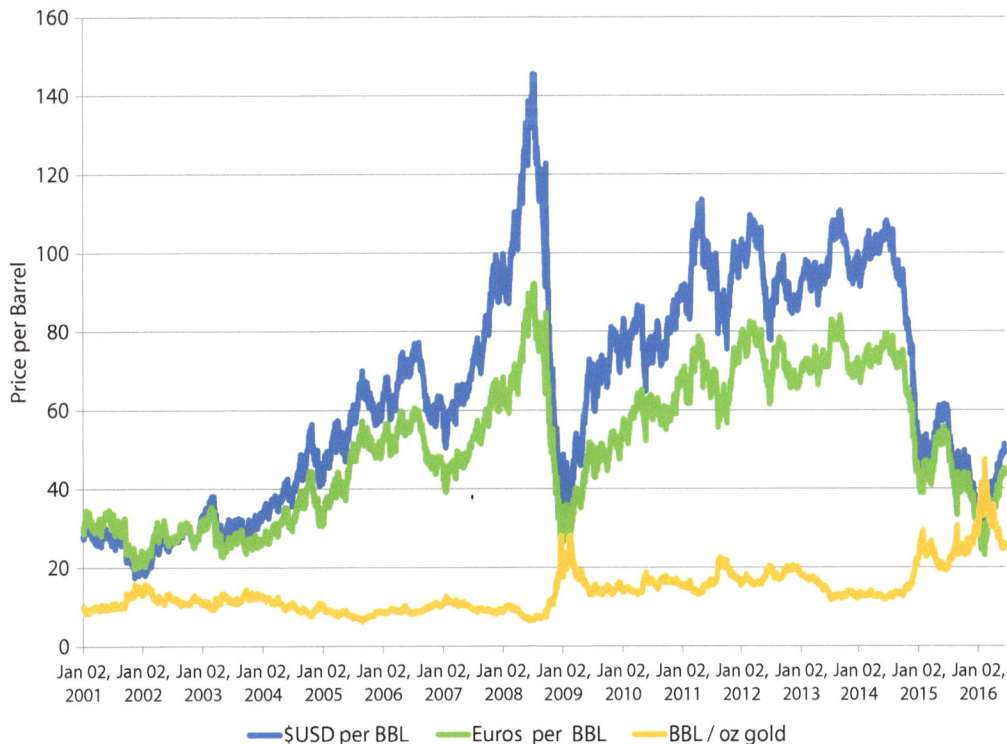

AGI Geoscience Workforce Program; Data derived from EIA, OANDA, and World Gold Council

Figure 5.6: Commodity Output for the U.S. Mining, Oil and Gas Extraction, and Support Industries

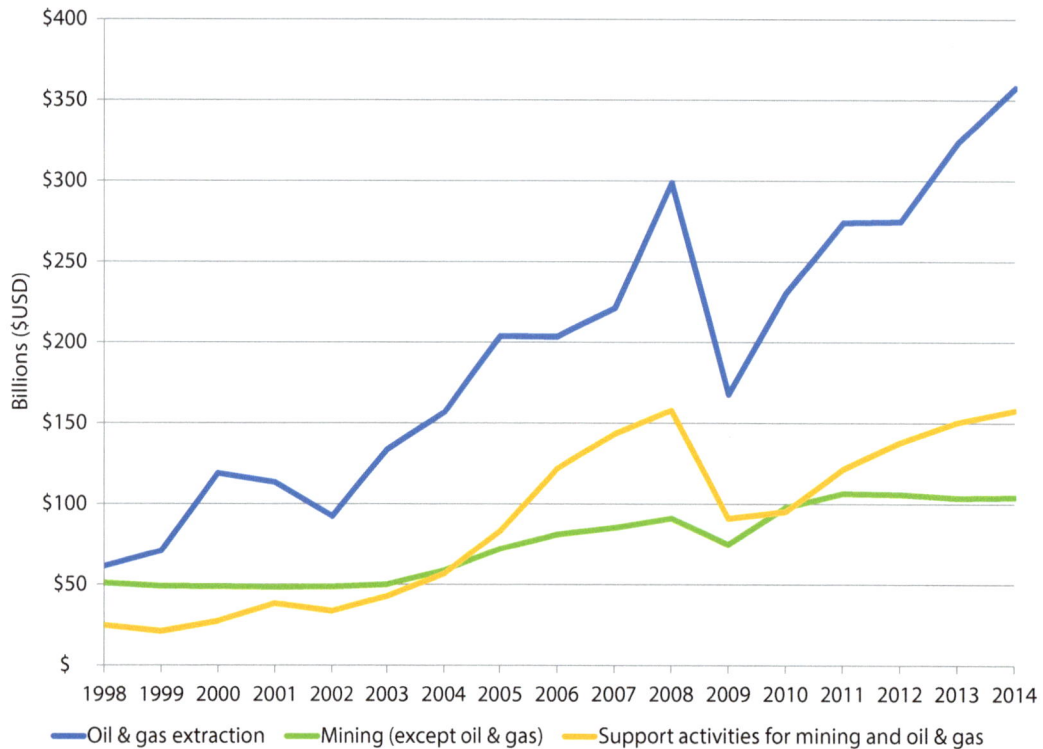

AGI Geoscience Workforce Program; Data derived from the U.S. Bureau of Economic Analysis

Figure 5.7: Gross Operating Surplus for the U.S. Mining, Oil and Gas Extraction, and Support Industries

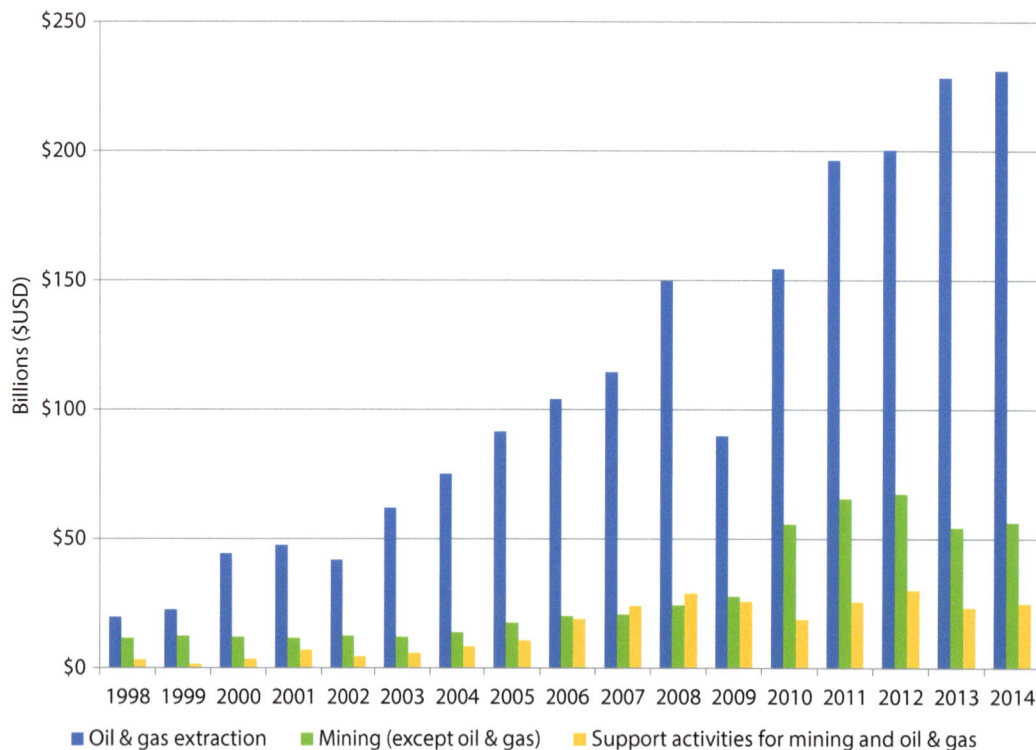

AGI Geoscience Workforce Program; Data derived from the U.S. Bureau of Economic Analysis

Gross Domestic Product Contribution of Geosciences

The geoscience component of industry gross domestic product (GDP) represents the first order economic contribution of geoscientists to the U.S. economy. The geoscience component of industry GDP is calculated by multiplying the value added amount for a specific industry by the percentage of the industry's total employment that are geoscientists. Thus, the total geoscience component of industry GDP is usually less than an industry's domestic production. For example, for the oil and gas industry, the value added amount was $302 billion in 2014. Geoscientists comprise 14.3% of the industry's employment. Therefore, the geoscience component of the oil and gas industry's GDP in 2012 was $43.78 billion (Figure 5.9). This was a decrease of approximately $10 billion since 2012 due to the decrease in the percentage of geoscientists in the petroleum industry compared to the total employment in the petroleum industry. 2014 was the beginning of the economic downturn in this industry, which included layoffs within the industry, and this can be seen in the change in the calculated GDP contribution due to geoscientists. Because the predicted increase in employment in the oil and gas industry for 2024 was less than the previous prediction for 2022, the predicted increase in GDP contribution from geoscientists is quite a bit lower than the prediction presented in the 2014 report.

Figure 5.8: Amount of Geoscience Industry GDP Contributed by Specific Industries

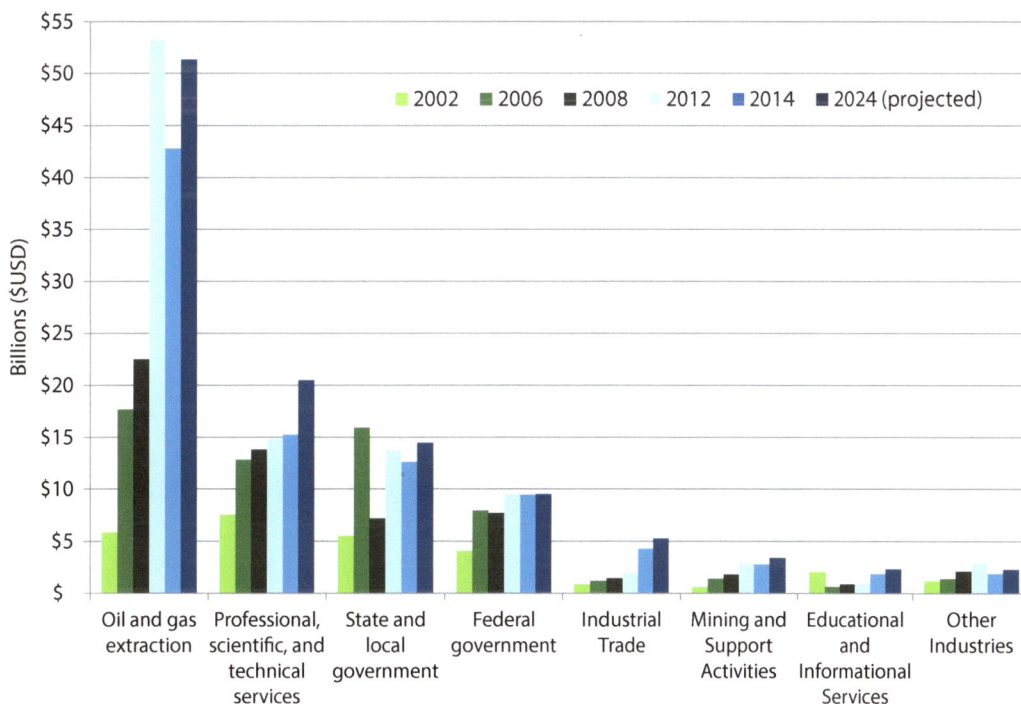

AGI Geoscience Workforce Program; Data derived from U.S. Bureau of Economic Analysis, U.S. Bureau of Labor Statistics, and AGI's Directory of Geoscience Departments database

Productive Activity of Geoscience Industries

At the height of the petroleum industry in 2014, 52% of the drilling rigs in the world were operating in the U.S., but at the end of 2015, that percentage dropped to 36% (Figure 5.9). The downturn in the petroleum industry over the past couple of years led to the deactivation of approximately 1,211 rigs since November of 2014. The majority of new U.S. wells continue to be onshore (land), but most of the idled rigs were on shore rigs (Figure 5.11). Since the recession in 2009, there has been a rapid increase in crude oil wells drilled and a rapid decrease in new natural gas wells. However, the past couple of years have shown a rapid decrease in crude oil rigs (Figure 5.13). The data was not updated for the average depth of wells drilled, so this figure is the same as the one that appeared in the 2014 report.

U.S. mines for industrial minerals and sand, gravel and stone appear to have been harder hit during the recession. While it appears industrial minerals mines have bounced back some in 2012, sand, gravel, and stone operations saw decreases in the material handled by the mines (Figure 5.15). In 2009, with the drop in sand, gravel, and stone mines, metal ore became the material with the highest yield in metric tons since 2009. Along with higher yields of metal ore, the value of the metal ore has risen sharply since 2009 reaching $35 billion in 2011-2012 (Figure 5.17).

Figure 5.9: Average Rotary Rig Counts by World Region

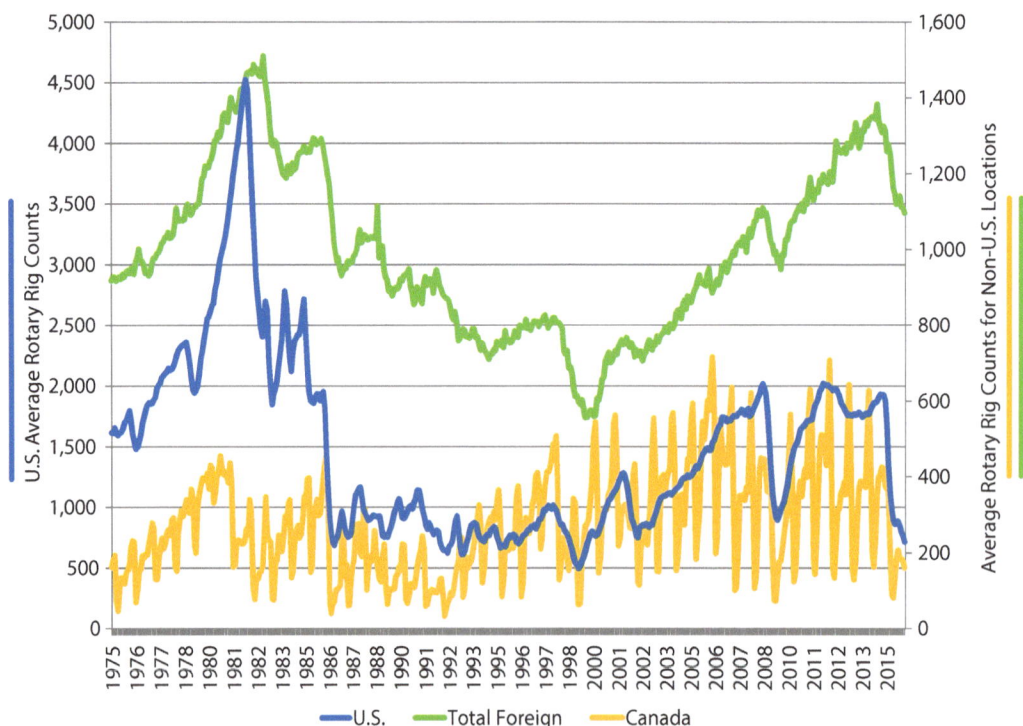

AGI Geoscience Workforce Program; Data derived from Baker Hughes

Figure 5.10: U.S. Rotary Rig Counts

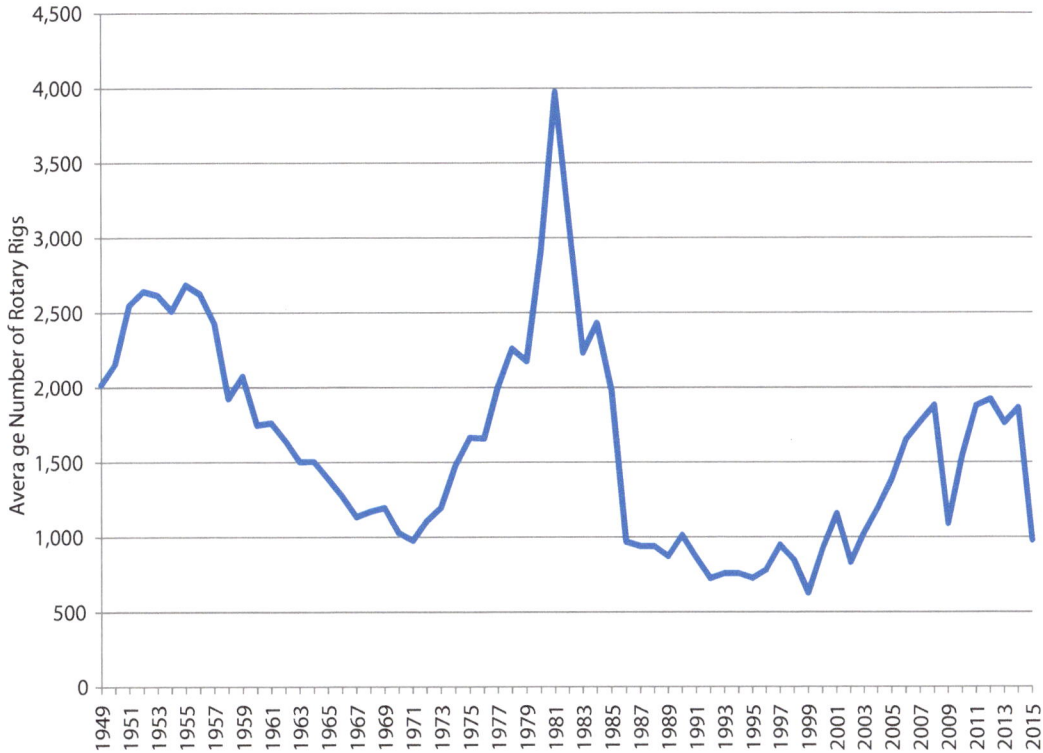

AGI Geoscience Workforce Program; Data derived from Baker Hughes

Figure 5.11: U.S. Rotary Rigs by Location

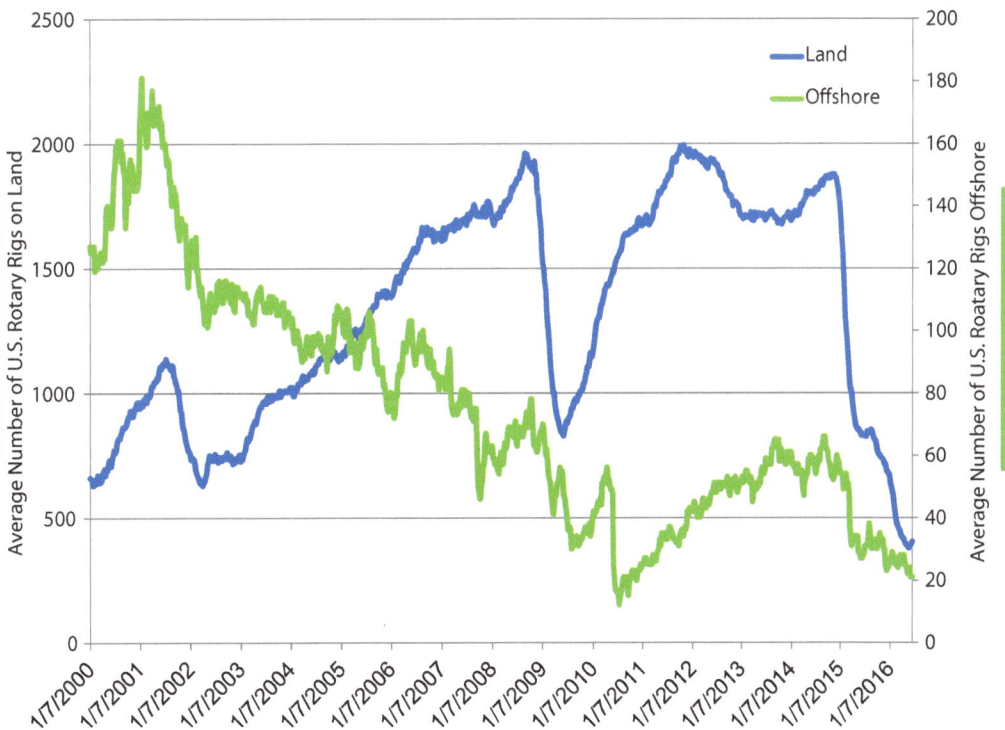

AGI Geoscience Workforce Program; Data derived from Baker Hughes

Figure 5.12: U.S. Rigs by Type

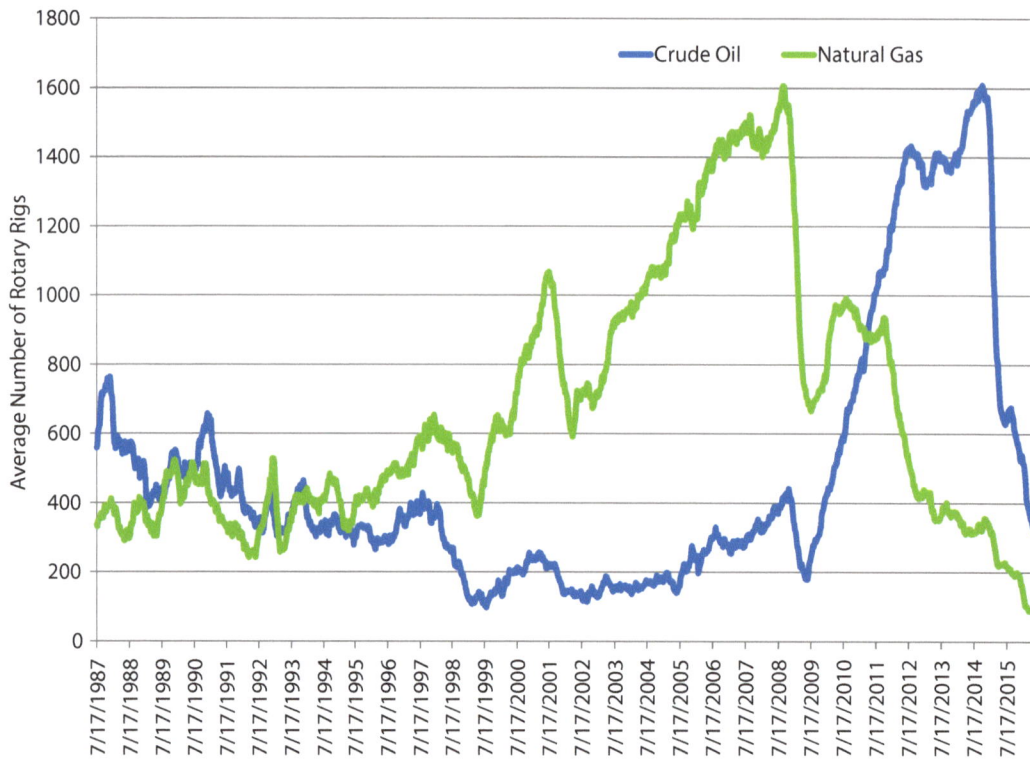

AGI Geoscience Workforce Program; Data derived from Baker Hughes

Figure 5.13: Average Depth of Wells Drilled by Type

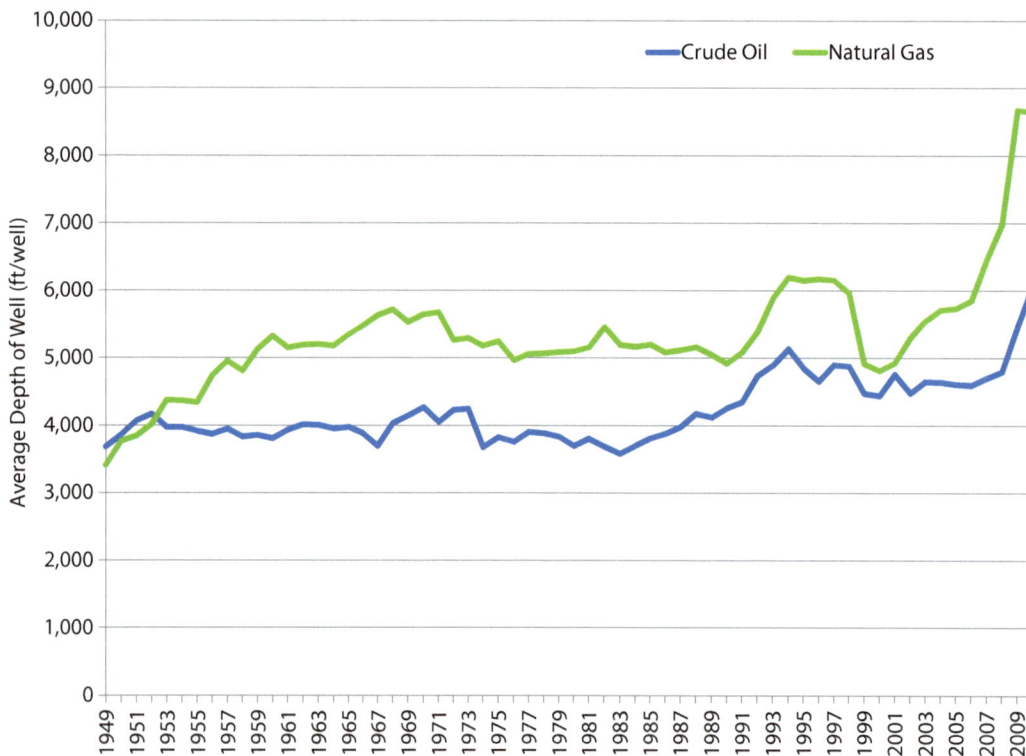

AGI Geoscience Workforce Program; Data derived from U.S. Energy Information Administration

Figure 5.14: Number of U.S. Mines

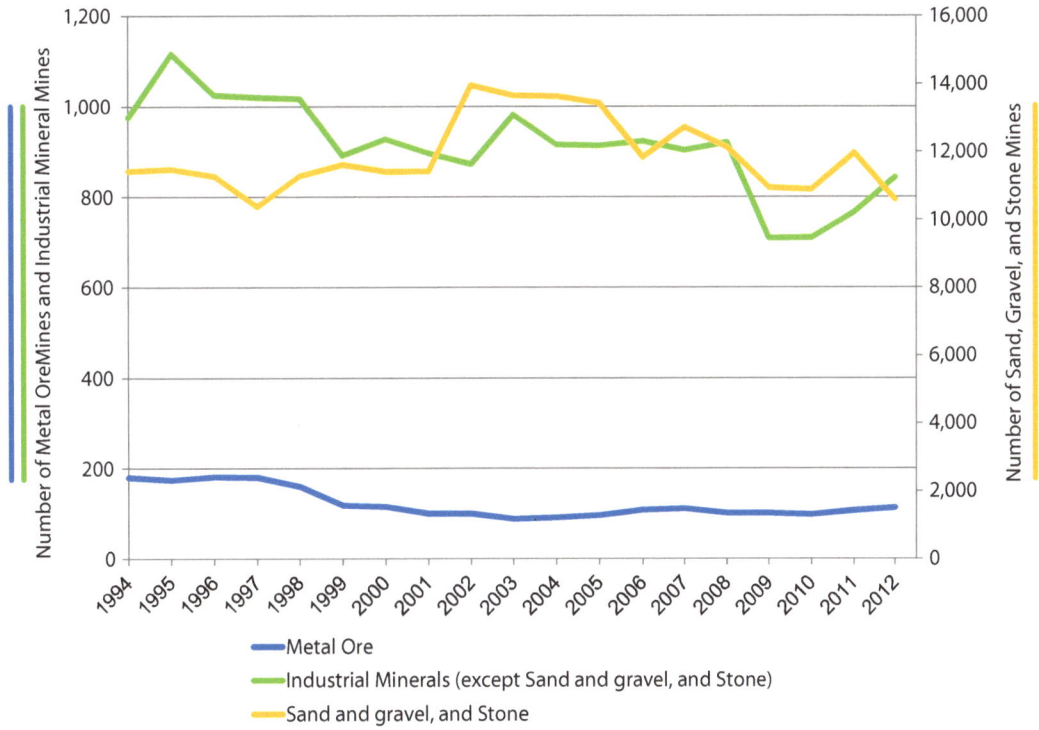

AGI Geoscience Workforce Program; Data derived from the USGS Mining and Quarrying Trends

Figure 5.15: Material Handled at U.S. Mines

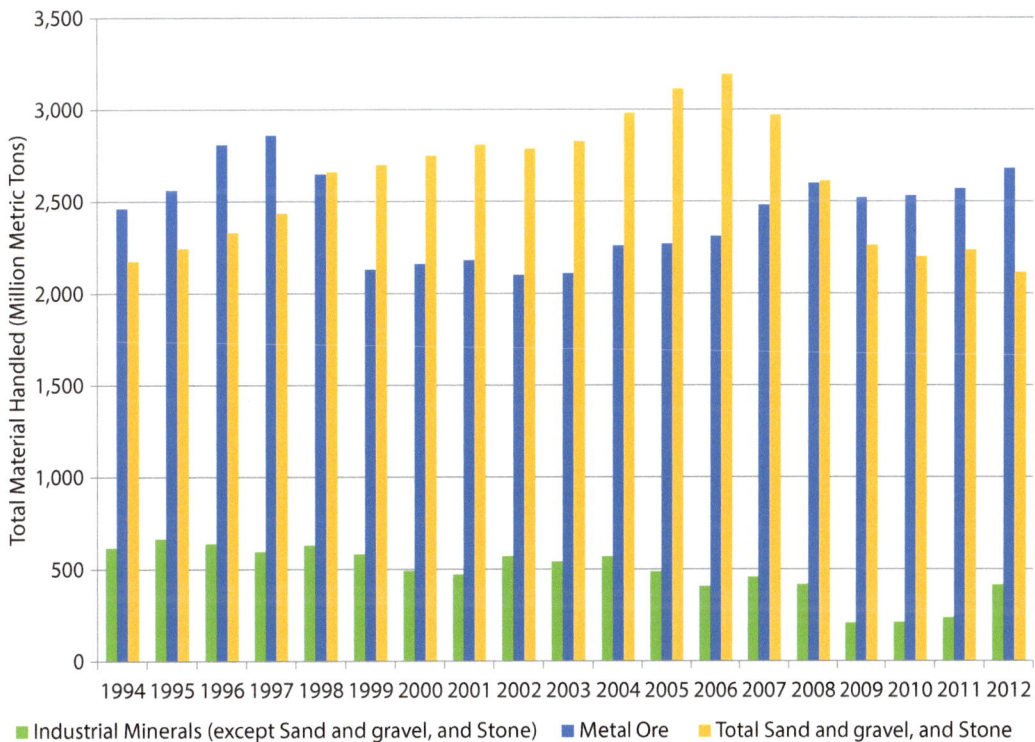

AGI Geoscience Workforce Program; Data derived from the USGS Mining and Quarrying Trends

Figure 5.16: Value of Non-Fuel Mineral Production from U.S. Mines

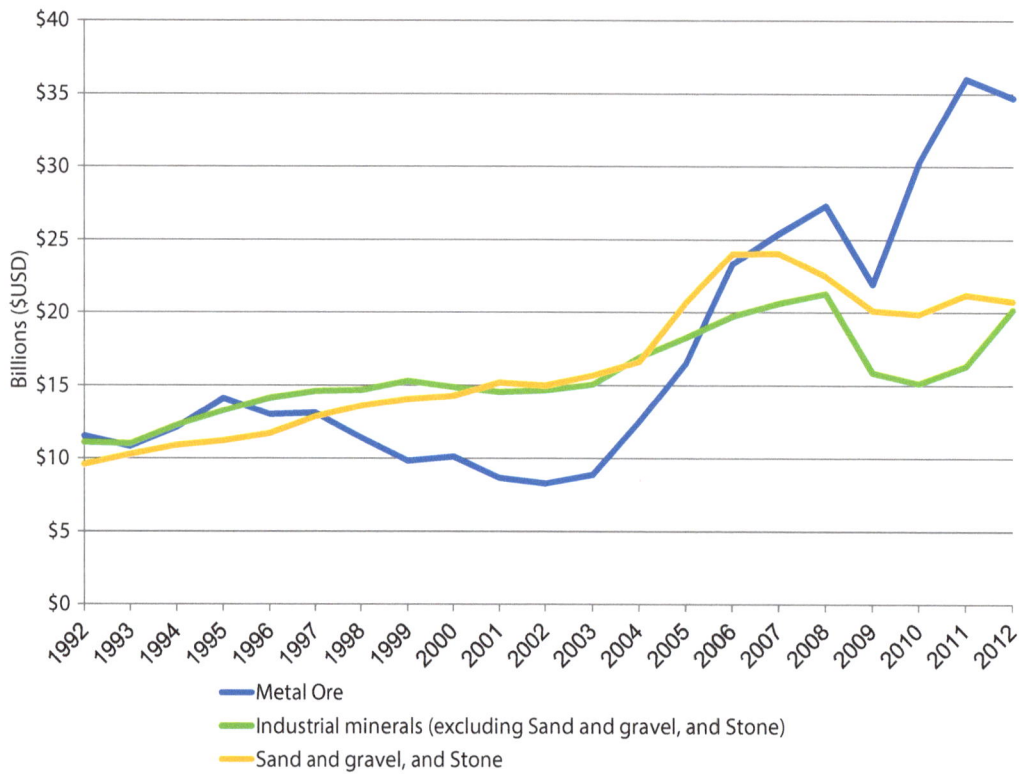

AGI Geoscience Workforce Program; Data derived from USGS Minerals Yearbook

Appendix A: Defining the Geosciences

Given its complexity, the geoscience occupation is difficult to define under existing nomenclature. This is the result of varied educational pathways geoscientists pursue and because of the different industries in which geoscientists work. Additionally, each federal data source (U.S. Bureau of Labor Statistics, U.S. Census Bureau, National Center for Education Statistics, National Science Foundation, U.S. Bureau of Economic Analysis, Office of Personnel Management), professional society, and industry classifies geoscientists differently depending on the intent of data collection (national occupational trends, science and engineering trends, education vs. occupation, internal classification codes, etc.), the characteristics of the population surveyed, and the focus of the organization.

U.S. federal policy and funding is partially determined by the economic activity and employment trends of a given profession. Accurate measurement and analysis of the geoscience profession are central to successful decisions that support a robust geoscience profession in the U.S.

Unfortunately, the geosciences are not consistently defined across the myriad of data sources collected and used by federal government and professional societies. In many cases, the issues of definition are related to splitting of disciplines. In some cases, they are archaic artifacts of early labor policy and, in other, represent a lack of domain knowledge in the agencies setting the definitions. Though many federal agencies are attempting to improve their classification approach, the current diversity of definitions will continue for the foreseeable future. Unfortunately, the public statistics from this data are used by counselors and individuals seeking career options, and the current state of geoscience workforce data tend to severely under-represents the size of the profession and the breadth of opportunities.

To address this issue, AGI has established a working definition for the geoscience profession in order to improve comparability of data across sources and time periods, which is laid out in this section.

Many federal data sources use the Classification Instructional Programs (CIP) codes to classify educational programs, the Standard Occupational Classification (SOC) codes to classify occupations, and the North American Industry Classification System (NAICS) to classify industries. In this appendix we report how each data source defines a geoscientist. The CIP codes are managed by the U.S. Department of Education's National Center for Education Statistics. The SOC codes were developed by the U.S. Office of Management and Budget and are managed by the Standard Occupational Classification Revision Policy Committee. This committee consists of representatives from the U.S. Bureau of Labor Statistics, the U.S. Bureau of Census, the U.S. Department of Labor (Employment and Training Administration), the Office of Personnel Management, the Defense Manpower Data Center, the National Science Foundation, the National Occupational Information Coordinating Committee, and the Office of Management and Budget. The NAICS was developed under the guidance of the Office of Management and Budget by the U.S. Economic Classification Policy Committee, Statistics Canada, and Mexico's Instituto Nacional de Estadistica, Geografia e Informatica in order to allow for economic comparisons between North American countries.

Educational Classifications

Classification of Instructional Programs (CIP)

The National Science Foundation and the National Center for Education Statistics use the Classification of Instructional Programs (CIP) to classify educational programs including fields of study and program completions.

The CIP website (https://nces.ed.gov/pubs2002/cip2000/) also has an online application that allows for the cross-referencing of instructional programs to the Standard Occupational Classification codes.

Appendix Table 1: CIP Codes that Refer to Geoscience Programs

CIP Code	Title	Description
3.0104	Environmental Science	A program that focuses on the application of biological, chemical, and physical principles to the study of the physical environment and the solution of environmental problems, including subjects such as abating or controlling environmental pollution and degradation; the interaction between human society and the natural environment; and natural resources management. Including instruction in biology, chemistry, physics, geosciences, climatology, statistics, and mathematical modeling.
14.0802	Geotechnical Engineering	A program that prepares individuals to apply mathematical and scientific principles to the design, development, and operational evaluation of systems for manipulating and controlling surface and subsurface features at or incorporated into structural sites, including earth and rock moving and stabilization, land fills, structural use and environmental stabilization of wastes and by-products, underground construction, and groundwater and hazardous material containment.
14.1401	Environmental/ Environmental Health Engineering	A program that prepares individuals to apply mathematical and scientific principles to the design, development and operational evaluation of systems for controlling contained living environments and for monitoring and controlling factors in the external natural environment, including pollution control, waste and hazardous material disposal, heath and safety protection, conservation, life support, and requirements for protection of special materials and related work environments.
14.2101	Mining and Mineral Engineering	A program that prepares individuals to apply mathematical and scientific principles to the design, development and operational evaluation of mineral extraction, processing and refining systems, including open pit and shaft mines, prospecting and site analysis equipment and instruments, environmental and safety systems, mine equipment and facilities, mineral processing and refining methods and systems, and logistics and communication systems.
14.2401	Ocean Engineering	A Program that prepares individuals to apply mathematical and scientific principles to the design, development, and operational evaluation of systems to monitor, control, manipulate, and operate within coastal or ocean environments, such as underwater platforms, flood control systems, dikes, hydroelectric power systems, tide and current control and warning systems, and communications equipment; the planning and design of total systems for working and functioning in water or underwater environments; and the analysis of related engineering problems such as the action of water properties and behavior on physical systems and people, tidal forces, current movements, and wave motion.
14.2501	Petroleum Engineering	A program that prepares individuals to apply mathematical and scientific principles to the design, development, and operational evaluation of systems for locating, extracting, processing and refining crude petroleum and natural gas, including prospecting instruments and equipment, mining and drilling systems, processing and refining systems and facilities, storage facilities, transportation systems, and related environmental and safety systems.
14.3901	Geological/ Geophysical Engineering	A program that prepares individuals to apply mathematical and geological principles to the analysis and evaluation of engineering problems, including the geological evaluation of construction sites, the analysis of geological forces acting on structures and systems, the analysis of potential natural resource recovery sites, and applied research on geological phenomena.
26.1302	Marine Biology and Biological Oceanography	A program that focuses on the scientific study of the ecology and behavior of microbes, plants, and animals inhabiting oceans, coastal waters, and saltwater wetlands and their interactions with the physical environment. Includes instruction in chemical, physical, and geological oceanography; molecular, cellular, and biochemical studies; marine microbiology; marine botany; ichthyology; mammalogy; marine population dynamics and biodiversity; reproductive biology; studies of specific species, phyla, habitats, and ecosystems; marine paleoecology and paleontology; and applications to fields such as fisheries science and biotechnology.

<No intersecting link>

CIP Code	Title	Description
40.0401	Atmospheric Sciences and Meteorology, General	A general program that focuses on the scientific study of the composition and behavior of the atmospheric envelopes surrounding the earth, the effect of earth's atmosphere on terrestrial weather, and related problems of environment and climate. Includes instruction in atmospheric chemistry and physics, atmospheric dynamics, climatology and climate change, weather simulation, weather forecasting, climate modeling and mathematical theory; and studies of specific phenomena such as clouds, weather systems, storms, and precipitation patterns.
40.0402	Atmospheric Chemistry and Climatology	A program that focuses on the scientific study of atmospheric constituents, reactions, measurement techniques, and processes in predictive, current, and historical contexts. Includes instruction in climate modeling, gases and aerosols, trace gases, aqueous phase chemistry, sinks, transport mechanisms, computer measurement, climate variability, paleoclimatology, climate diagnosis, numerical modeling and data analysis, ionization, recombination, photoemission, and plasma chemistry
40.0403	Atmospheric Physics and Dynamics	A program that focuses on the scientific study of the processes governing the interactions, movement, and behavior of atmospheric phenomena and related terrestrial and solar phenomena. Includes instruction in cloud and precipitation physics, solar radiation transfer, active and passive remote sensing, atmospheric electricity and acoustics, atmospheric wave phenomena, turbulence and boundary layers, solar wind, geomagnetic storms, coupling, natural plasma, and energization.
40.0404	Meteorology	A program that focuses on the scientific study of the prediction of atmospheric motion and climate change. Includes instruction in general circulation patterns, weather phenomena, atmospheric predictability, parameterization, numerical and statistical analysis, large-and mesoscale phenomena, kinematic structures, precipitation processes, and forecasting techniques.
40.0499	Atmospheric Sciences and Meteorology, Other	Any instructional program in atmospheric sciences and meteorology not listed above.
40.0601	Geology/ Earth Sciences, General	A program that focuses on the scientific study of the earth; the forces acting upon it; and the behavior of the solids, liquids and gases comprising it. Includes instruction in historical geology, geomorphology and sedimentology, the chemistry of rocks and soils, stratigraphy, mineralogy, petrology, geostatistics, volcanology, glaciology, geophysical principles, and applications to research and industrial problems.
40.0602	Geochemistry	A program that focuses on the scientific study of the chemical properties and behavior of the silicates and other substances forming, and formed by geomorphological processes of the earth and other planets. Includes instruction in chemical thermodynamics, equilibrium in silicate systems, atomic bonding, isotopic fractionation, geochemical modeling, specimen analysis, and studies of specific organic and inorganic substances.
40.0603	Geophysics and Seismology	A program that focuses on the scientific study of the physics of solids and its application to the study of the earth and other planets. Includes instruction in gravimetric, seismology, earthquake forecasting, magnetometry, electrical properties of solid bodies, plate tectonics, active deformation, thermodynamics, remote sensing, geodesy, and laboratory simulations of geological processes.
40.0604	Paleontology	A program that focuses on the scientific study of extinct life forms and associated fossil remains, and the reconstruction and analysis of ancient forms, ecosystems, and geological processes. Includes instruction in sedimentation and fossilization processes, fossil chemistry, evolutionary biology, paleoecology, paleoclimatology, trace fossils, micropaleontology, invertebrate paleontology, vertebrate paleontology, paleobotany, field research methods, and laboratory research and conservation methods.
40.0605	Hydrology and Water Resources Science	A program that focuses on the scientific study of the occurrence, circulation, distribution, chemical and physical properties, and environmental interaction of surface and subsurface waters, including groundwater. Includes instruction in geophysics, thermodynamics, fluid mechanics, chemical physics, geomorphology, mathematical modeling, hydrologic analysis, continental water processes, global water balance, and environmental science.
40.0606	Geochemistry and Petrology	A program that focuses on the scientific study of the igneous, metamorphic, and hydrothermal processes within the earth and the mineral, fluid, rock, and ore deposits resulting from them. Includes instruction in mineralogy, crystallography, petrology, volcanology, economic geology, meteoritics, geochemical reactions, deposition, compound transformation, core studies, theoretical geochemistry, computer applications, and laboratory studies.
40.0607	Oceanography, Chemical and Physical	A program that focuses on the scientific study of the chemical components, mechanisms, structure, and movement of ocean waters and their interaction with terrestrial and atmospheric phenomena. Includes instruction in material inputs and outputs, chemical and biochemical transformations in marine systems, equilibria studies, inorganic and organic ocean chemistry, oceanographic processes, sediment transport, zone processes, circulation, mixing, tidal movements, wave properties, and seawater properties.
40.0699	Geological and Earth Sciences/ Geosciences, Other	Any instructional program in geological and related sciences not listed above.
45.0701	Geography	A program that focuses on systematic study of the spatial distribution and interrelationships of people, natural resources, plant and animal life. Includes instruction in historical and political geography, cultural geography, economic and physical geography, regional science, cartographic methods, remote sensing, spatial analysis, and applications to areas such as land-use planning, development studies, and analyses of specific countries, regions, and resources.

Occupational Classifications

Standard Occupational Classification Codes

The U.S. Census Bureau of Labor Statistics and National Science Foundation (NSF) use the 2010 Standard Occupational Classification (SOC) codes (http://www.bls.gov/soc/) to classify geoscientists; however, each organization has a different focus for its surveying and data collection.

Data from the U.S. Census Bureau, U.S. Bureau of Labor Statistics and the Office of Personnel Management are coarse because the first two agencies focus on national population trends and the third agency focuses on trends across all sectors of the federal government. Data from the National Science Foundation has a finer resolution because it is focused on specific data topics within the science and engineering fields. Data from all of these sources are too coarse to establish precise trends for geoscientists.

In data classified by the SOC codes, some geoscientists are grouped in categories with other non-geoscience scientists and engineers. For example, soil scientists who study the chemical, physical, and mineralogical composition of soils are grouped with the Soil and Plant Scientists whose focus is on agriculture. Geotechnical engineers, who study the structural behavior of soil and rocks, perform soil investigations, design structure foundations, and provide field observations of foundation investigation and construction, are grouped with Civil Engineers who perform construction. Geoscientists at the professional or managerial level are grouped with either Engineering Managers or Natural Science Managers. Geoscience teachers at post-secondary institutions are grouped in the Environmental Science Teacher, Atmospheric, Earth, Marine, and Space Science Teacher, Geography Teacher, or Engineering Teacher categories.

The National Science Foundation's classification of geoscientists provides better resolution than the SOC codes; however, there are no categories for geographers, hydrologists, geoscience managers and soil scientists. Additionally, many of the challenges with identifying geoscientists that occur in the SOC codes (such as post-secondary geoscience teachers) also occur within the National Science Foundation's classification schema.

Appendix Table 2: Geoscientists are Found within the Following SOC Codes

SOC Code	SOC Title	Definition
11-9041	Architectural and Engineering Managers	Plan, direct, or coordinate activities in such fields are architecture and engineering or research and development in these fields. Excludes "Natural Sciences Managers"
11-9121	Natural Science Managers	Plan, direct, or coordinate activities in such fields as life sciences, physical sciences, mathematics, statistics, and research and development in these fields. Excludes "Architectural and Engineering Managers" and "Computer and Information Systems Managers"
17-2051	Civil Engineers	Perform engineering duties in planning, designing, and overseeing construction and maintenance of building structures and facilities, such as roads, railroads, airports, bridges, harbors, channels, dams, irrigation projects, pipelines, power plants, and water and sewage systems. Includes architectural, structural, traffic, ocean, and geo-technical engineers. Excludes "Hydrologists".
17-2081	Environmental Engineers	Research, design, plan or perform engineering duties in the prevention, control, and remediation of environmental hazards using various engineering disciplines. Work may include waste treatment, site remediation, or pollution control technology.
17-2151	Mining and Geological Engineers, Including Mining Safety Engineers	Conduct sub-surface surveys to identify the characteristics of potential land or mining development sites. May specify the ground support systems, processes and equipment for safe, economical, and environmentally sound extraction or underground construction activities. May inspect areas for unsafe geological conditions, equipment, and working conditions. May design, implement, and coordinate mine safety programs. Excludes "Petroleum Engineers".
17-2171	Petroleum Engineers	Devise methods to improve oil and gas extraction and production and determine the need for new or modified tool designs. Oversee drilling and offer technical advice.

SOC Code	SOC Title	Definition
19-1013	Soil and Plant Scientists	Conduct research in breeding, physiology, production, yield, and management of crops and agricultural plants or trees, shrubs. and nursery stock, their growth in soils, and control of pests; or study the chemical, physical, biological, and mineralogical composition of soils as they relate to plant or crop growth. May classify and map soils and investigate effects of alternative practices on soil and crop productivity.
19-1031	Conservation Scientists	Manage, improve and protect natural resources to maximize their use without damaging the environment. May conduct soil surveys and develop plans to eliminate soil erosion or to protect rangelands. May instruct farmers, agricultural production managers, or ranchers in best ways to use crop rotation, contour plowing, or terracing to conserve soil and water; in the number and kind of livestock and forage plants best suited to particular ranges; and in range and farm improvements, such as fencing and reservoirs for stock watering. Excludes "Zoologists and Wildlife Biologists" and "Foresters"
19-2021	Atmospheric and Space Scientists	Investigate atmospheric phenomena and interpret meteorological data, gathered by surface and air stations, satellites, and radar to prepare reports and forecasts for public and other data uses. Includes weather analysts and forecasters whose functions require the detailed knowledge of meteorology.
19-2041	Environmental Scientists and Specialists, Including Health	Conduct research or perform investigation for the purpose of identifying, abating, or eliminating sources of pollutants of hazards that affect either the environment or the health of the population. Using knowledge of various scientific disciplines, may collect, synthesize, study, report, and recommend action based on data derived from measurements or observations of air, food, soil, water, and other sources. Excludes "Zoologists and Wildlife Biologists", "Conservation Scientists", "Forest and Conservation Technicians", "Fish and Game Wardens", and "Forest and Conservation Workers".
19-2042	Geoscientists, Except Hydrologists and Geographers	Study the composition, structure, and other physical aspects of the Earth. May use geological, physics, and mathematics knowledge in exploration for oil, gas, minerals, or underground water; or in waste disposal, land reclamation, or other environmental problems. May study the Earth's internal composition, atmospheres, oceans, and its magnetic, electrical, and gravitational forces. Includes mineralogists, crystallographers, paleontologists, stratigraphers, geodesists, and seismologists.
19-2043	Hydrologists	Research the distribution, circulation, and physical properties of underground and surface waters; and study the form and intensity of precipitation, its rate of infiltration into the soil, movement through the earth, and its return to the ocean and atmosphere.
19-3092	Geographers	Study the nature and use of areas of the Earth's surface, relating and interpreting interactions of physical and cultural phenomena. Conduct research on physical aspects of a region, including land forms, climates, soils, plants, and animals, and conduct research on the spatial implications of human activities within a given area, including social characteristics, economic activities, and political organization, as well as researching interde-pendence between regions at scales ranging from local to global.
19-4041	Geological and Petroleum Technicians	Assist scientists or engineers in the use of electronic, sonic, or nuclear measuring instruments in both laboratory and production activities to obtain data indicating potential resources such as metallic ore, minerals, gas, coal, or petroleum. Analyze mud and drill cuttings. Chart pressure, temperature, and other characteristics of wells or bore holes. Investigate and collect information leading to the possible discovery of new metallic ore, minerals, gas, coal, or petroleum deposits.
19-4091	Environmental Science and Protection Technicians, Including Health	Perform laboratory and field tests to monitor the environment and investigate sources of pollution, including those that affect health, under the direction of an environmental scientist, engineer, or other specialist. May collect samples of gases, soil, water, and other materials for testing.
25-1032	Engineering Teachers, Postsecondary	Teach courses pertaining to the application of physical laws and principles of engineering for the development of machines, materials, instruments, processes, and services. Includes teachers of subjects such as chemical, civil, electrical, industrial, mechanical, mineral, and petroleum engineering. Includes both teachers primarily engaged in teaching and those who do a combination of teaching and research. Excludes "Computer Science Teachers, Postsecondary".
25-1051	Atmospheric. Earth, Marine, and Space Sciences Teachers, Postsecondary	Teach courses in the physical sciences, except chemistry and physics. Includes both teachers primarily engaged in teaching, and those who do a combination of teaching and research.
25-1053	Environmental Science Teachers, Postsecondary	Teach courses in environmental science. Includes both teachers primarily engaged in teaching and those who do a combination of teaching and research.
25-1064	Geography Teachers, Postsecondary	Teach courses in geography. Includes both teachers primarily engaged in teaching and those who do a com-bination of teaching and research.

Office of Personnel Management: Handbook of Occupations Groups and Families

The Office of Personnel Management released this Handbook in order to provide agencies with a starting point to classify positions.

Appendix Table 3: Geoscientists are Found within the Following OPM Handbook Codes

Code-Title	Description
0028-Environmental Protection Specialist Series	This series covers positions that involve advising on, managing, supervising, or performing administrative or program work relating to environmental protection programs (e.g., programs to protect or improve environmental quality, control pollution, remedy environmental damage, or ensure compliance with environmental laws and regulations). These positions require specialized knowledge of the principles and methods of administering environmental protection programs and the laws and regulations related to environmental protection activities.
0150-Geography Series	This series covers positions the duties of which involve professional work in the field of geography, including the compilation, synthesis, analysis, interpretation and presentation of information regarding the location, distribution, and interrelationships of and processes of change affecting such natural and human phenomena as the physical features of the earth, climate, plant, and animal life, and human settlements and institutions.
0401-General Natural Resources Management and Biological Science Series	This series covers positions that involve professional work in biology, agriculture, or related natural resource management when there is no other more appropriate series. Thus included in this series are positions that involve: 1) a combination of several professional fields with none predominant; or 2) a specialized professional field not readily identified with other existing series.
0457-Soil Conservation Series	This series covers positions involving the performance of professional work in the conservation of soil, water, and related environmental resources to achieve sound land use. Conservation work requires knowledge of: 1) soils and crops; 2) the pertinent elements of agronomy, engineering, hydrology, range conservation, biology, and forestry; and 3) skill in oral and written communication methods and techniques sufficient to impart these knowledge to selected client groups.
0470-Soil Science Series	This series covers positions that involve professional and scientific work in the investigation of soils, their management, and their adaptation for alternative uses. Such work requires knowledge of chemical, physical, mineralogical and biological properties and processes of the soils and their relationships to climatic, physiographic, and biologic influences.
0819-Environmental Engineering Series	This series covers positions managing, supervising, leading, and/or performing professional engineering and scientific work involving environmental programs and projects in the areas of: 1) environmental planning; 2) environmental compliance; 3) identification and cleanup of contamination; and 4) restoring and sustaining environmental conservation.
0880-Mining Engineering Series	This series covers positions managing, supervising, leading, and/or performing professional engineering and scientific work to explore, remove, and transport raw metals, nonmetallic minerals, and solid fuels from the earth. Mining engineering work involves: 1) a variety of mineral substances to include metal ores, nonmetallic minerals, and solid fuels and energy sources; 2) working with mining systems, including underground mining, surface mining, solution mining, and placer mining; and 3) traditional mining activities, including the heavy construction industry (involving rock excavation and support for highways, tunnels, dams, power stations, and underground chambers) and exploration and development of mineral deposits located under large bodies of water.
0881-Petroleum Engineering Series	This series covers positions managing, supervising, leading, and/or performing professional engineering and scientific work involved in the discovery and recovery of oil, natural gas (e.g. methane, ethane, propane, butane), and helium. The work includes: 1) exploration and development of oil and natural gas fields; 2) production, transportation, and storage of petroleum, natural gas, and helium; 3) investigation, evaluation, and conservation of these resources; 4) regulation of the transportation and sale of natural gas; 5) valuation of production and distribution facilities for tax, regulatory, and other purposes; and 6) research on criteria, principles. methods, and equipment involved in exploration and development activities.
1301-General Physical Science Series	This series includes positions that involve professional work in the physical sciences when there is no other more appropriate series, that is, the positions are not classifiable elsewhere. This series also includes work in a combination of physical science fields, with no one predominant.
1313-Geophysics Series	This series includes professional scientific positions requiring application of knowledge of the principles and techniques of geophysics and related sciences in the investigation, measurement, analysis, evaluation, and interpretation of geophysical phenomena and artificially applied forces and fields related to the structure, composition, and physical properties of the earth and its atmosphere.
1315-Hydrology Series	This series includes positions that involve professional work in hydrology, the science concerned with the study of water in the hydrologic cycle. The work includes basic and applied research on water and water resources; the collection, measurement, analysis, and interpretation of information on water resources; the forecast of water supply and water flows; and the development of new, improved or more economical methods, techniques, and instruments.
1321-Metallurgy Series	This series includes positions that require primarily professional education and training in the field of metallurgy, including ability to apply the relevant principles of chemistry, physics, mathematics, and engineering to the study of metals. Metallurgy is the art and science of extracting metals from their ores, refining them, alloying them and preparing them for use, and studying their properties and behavior as affected by the composition, treatment in manufacture, and condition of use.

Code-Title	Description
1340-Meteorology Series	This series includes positions that involve professional work in meteorology, the science concerned with the earth's atmospheric envelope and its processes. The work includes basic and applied research into the conditions and phenomena of the atmosphere; the collection, analysis, evaluation, and interpretation of meteorological data to predict weather and determine climatological conditions for specific geographical areas; the development of new improvements of existing meteorological theory; and the development or improvement of meteorological methods, techniques, and instruments. Positions in this occupation require full professional knowledge and application of meteorological methods, techniques, and theories.
1350-Geology Series	This series includes professional scientific positions applying a knowledge of the principles and theories of geology and related sciences in the collection, measurement, analysis, evaluation, and interpretation of geologic information concerning the structure, composition, and history of the earth. This includes the performance of basic research to establish fundamental principles and hypotheses to develop a fuller knowledge and understanding of geology, and the application of these principles and knowledge to a variety of scientific, engineering, and economic problems.
1360-Oceanography Series	This series includes professional scientific positions engaged in the collection, measurement, analysis, evaluation, and interpretation of natural and physical ocean phenomena, such as currents, circulations, waves, beach and near-shore processes, chemical structure and processes, physical and submarine features, depth, floor configuration, organic, and inorganic sediments, sound and light transmission, color manifestations, heat exchange, and similar phenomena (e.g. biota, weather, geological structure, etc.). Oceanographers plan, organize, conduct, and administer seagoing and land-based study and research of ocean phenomena for the purpose of interpreting, predicting, utilizing and controlling ocean forces and events, This work requires a fundamental background in chemistry, physics, mathematics, and appropriate knowledge in the field of oceanography.

Industry Classifications

North American Industry Classification System (NAICS)

The NAICS (https://www.census.gov/eos/www/naics/) is the federal government's standard industry classification system that groups employers into industries based on the activities in which they are primarily engaged. The United States, Canada, and Mexico developed the system to provide comparable statistics across the three countries. The NAICS is a comprehensive system covering the entire field of economic activities. There are 20 sectors in the NAICS and 1,065 detailed industries in the NAICS for the United States. The NAICS (United States version) is used by U.S. statistical agencies to facilitate the collection, tabulation, presentation, and analysis of data relating to business establishments. It allows for interagency comparison of statistical data describing the U.S. economy. The NAICS is used by the U.S. Census Bureau, U.S. Bureau of Labor Statistics, U.S. Bureau of Economic Analysis, and by the National Science Foundation.

The top-level categories for NAICS are outlined in following table. Geoscientists work in the Mining, Utilities, Construction, Manufacturing, Wholesale Trade, Transportation and Warehousing, Information, Finance and Insurance, Professional, Scientific, and Technical Services, Management of Companies and Enterprises, Administrative and Support and Waste Management and Remediation Services, Educational Services, and Public Administration industries.

Appendix Table 4: All Occupations Fall Within the Following NAICS Codes

NAICS Code	NAICS Industry Title
11	Agriculture, Forestry, Fishing and Hunting
21	Mining, Quarrying, and Oil and Gas Extraction
22	Utilities
23	Construction
31-33	Manufacturing
42	Wholesale Trade
44-45	Retail Trade
48-49	Transportation and Warehousing
51	Information
52	Finance and Insurance
53	Real Estate and Rental and Leasing
54	Professional, Scientific, and Technical Services
55	Management of Companies and Enterprises
56	Administrative and Support and Waste Management and Remediation Services
61	Educational Services
62	Health Care and Social Assistance
71	Arts, Entertainment, and Recreation
72	Accommodation and Food Services
81	Other Services (except Public Administration)
92	Public Administration

AGI's Working Definition of Geoscience Occupations

In light of how existing data sources define the geosciences, AGI has worked with its stakeholders to establish a working definition for the geoscience profession in order to improve compatibility of data across sources and time periods. With this definition, AGI and its partners will be able to capture the depth and breadth of the geoscience profession, clearly define it, and estimate employment trends. The resulting data can then be used in a proposal to federal data agencies to more accurately define the geosciences in federal data sources.

AGI's working definition of the geosciences is as follows:

Geoscientist

Subfields: Environmental Science, Hydrology, Oceanography, Atmospheric Science, Geology, Geophysics, Climate Science, Geochemistry, Paleontology

Studies the composition, structure, and other physical aspects of the Earth. Includes the study of the chemical, physical, and mineralogical composition of soils, analysis of atmospheric phenomena, and study the distribution, circulation, and physical and chemical properties of underground and surface waters. May study the Earth's internal composition, atmospheres, oceans, and its magnetic, electrical, thermal, and gravitational forces. May utilize knowledge of various scientific disciplines to collect, synthesize, study, report, and take action based on data derived from measurements or observations of air, soil, water, and other resources. May use geological, environmental, physics, and mathematics knowledge in exploration for oil, gas, minerals, or underground water; or in waste disposal, elimination of pollutants/hazards that affect the environment, land reclamation, or management of natural resources.

Geoscience Engineer

Subfield: Environmental

Designs, plans, or performs engineering duties in the development of water supplies and prevention, control, and remediation of environmental hazards utilizing various engineering disciplines. Work may include waste treatment, site remediation, pollution control technology, or the development of water supplies.

Subfield: Exploration

Determines the location and plans the extraction of coal, metallic ores, nonmetallic minerals, and building materials, such as stone and gravel. Work involves conducting preliminary surveys of deposits or undeveloped mines and planning their development; examining deposits or mines to determine whether they can be worked at a profit; making geological and topographical surveys; evolving methods of mining best suited to character, type, and size of deposits; and supervising mining operations. Devises methods to improve oil and gas well production and determines the need for new or modified tool designs. Oversees drilling and offers technical advice to achieve economical and satisfactory progress.

Subfield: Geotechnical

Studies structural behavior of soils and rocks, performs soil investigations, designs structure foundations, and provides field observations of foundation investigation and foundation construction.

Geoscience Manager

Plans, directs, or coordinates activities in such fields as geoscience engineering and geoscience. Engages in complex analysis of geoscience principles. Generally oversees one or more professionals, but may still be active in technical work.

Appendix B: Data Sources

AGI Data Sources:

GeoRef

AGI's GeoRef database contains over 4 million references to geoscience journal articles, books, maps, conference papers, reports and theses. GeoRef includes all geoscience publications that pertain only to surface and sub-surface processes. Publications pertaining to atmospheric and space sciences are excluded.

AGI's Directory of Geoscience Departments Database

AGI's Workforce Program has been collecting basic demographic information annually about all the geoscience programs at two-year and four-year institutions worldwide, along with other geoscience organizations and agencies for nearly 50 years, creating an extensive database. This database is used for the AGI publication, the Directory of Geoscience Departments, and for current data on the health of geoscience departments in the U.S. and abroad.

AGI's Geoscience Student Exit Survey

AGI collects data from students graduating with their bachelor's, master's, and doctoral geoscience degrees in order to ascertain their educational background, degree information, co-curricular experiences, and immediate future plans after graduation. Some of the data are presented within this report. For more information about this study and to see more of the most recent data, please visit http://www.americangeosciences.org/workforce/exit-survey.

Other Public Sources:

American Geophysical Union publications:
http://agupubs.onlinelibrary.wiley.com/

American Meteorological Society publications:
http://journals.ametsoc.org/

Association for the Sciences of Limnology and Oceanography publication:
http://www.aslo.org/lo/toc/index.html

ACT: https://www.act.org/

Baker Hughes: http://www.bakerhughes.com/

College Board: https://www.collegeboard.org/

Energy Information Administration:
http://www.eia.gov/

National Association of Geoscience Teachers publication: http://nagt.org/nagt/jge/index.html

National Science Foundation's Budget Internet Information Systems:
http://dellweb.bfa.nsf.gov/starth.asp

National Science Foundation's National Center for Science and Engineering Statistics:
http://www.nsf.gov/statistics/

U.S. Bureau of Economic Analysis:
http://www.bea.gov/

U.S. Bureau of Labor Statistics: http://www.bls.gov

U.S. Census Bureau: https://www.census.gov/

U.S. Department of Education's Integrated Postsecondary Education Database:
http://nces.ed.gov/ipeds/

U.S. Department of Education's National Center of Education Statistics: http://nces.ed.gov/

U.S. Geological Survey: http://www.usgs.gov

U.S. Government's Open Data Site:
http://www.data.gov/

U.S. Office of Personnel Management:
http://www.opm.gov/

World Gold Council: http://www.gold.org/

www.ingramcontent.com/pod-product-compliance
Lightning Source LLC
Chambersburg PA
CBHW041445210326
41599CB00004B/141